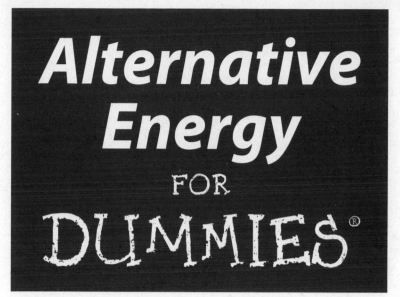

Alternative Energy FOR DUMMIES®

by Rik DeGunther

WILEY

Wiley Publishing, Inc.

Alternative Energy For Dummies®

Published by
Wiley Publishing, Inc.
111 River St.
Hoboken, NJ 07030-5774
www.wiley.com

Copyright © 2009 by Wiley Publishing, Inc., Indianapolis, Indiana

Published simultaneously in Canada

For general information on our other products and services, please contact our Customer Care Department within the U.S. at 877-762-2974, outside the U.S. at 317-572-3993, or fax 317-572-4002.

For technical support, please visit www.wiley.com/techsupport.

Wiley also publishes its books in a variety of electronic formats. Some content that appears in print may not be available in electronic books.

Library of Congress Control Number: 2009925036

ISBN: 978-0-470-43062-0

Manufactured in the United States of America

10 9 8 7 6 5 4 3 2

WILEY

About the Author

Rik DeGunther attended the University of Illinois as an undergraduate and Stanford University as a graduate student, studying both applied physics and engineering economics (some of this education actually stuck). He holds several United States patents and has designed a wide range of technical equipment including solar energy platforms, military-grade radar jammers, weather-measurement equipment, high-powered radar vacuum tubes, computerized production hardware, golf practice devices, digital and analog electronic circuits, unmanned aerial vehicles, guitars and amplifiers, microwave filters and mixers, automatic cabinet openers, strobe light communications systems, explosive devices (strictly on accident), cloud-height sensors, fog sensors, furniture, houses, barns, rocket ships, dart throwers, flame throwers, eavesdropping devices, escape routes, and you name it. He's one of those nerdy guys who likes to take things apart to see how they work and then put them back together and try to figure out what the leftover parts are for.

Rik is CEO of Efficient Homes, an energy-efficiency auditing firm in Northern California. He is actively engaged in designing and developing new solar equipment, including off-grid lighting systems and off-grid swimming pool heaters. He writes weekly op-ed columns for the Mountain Democrat, California's oldest and most venerable newspaper. He has also written a highly acclaimed golf book (on putting) and spends most of his free time attempting to improve his relatively impressive but objectively droll golf handicap, usually to no avail. Sometimes the urge strikes him to play a very loud guitar, of which he owns a collection with far more intrinsic quality than the playing they receive. His hearing has been faltering the last few years, so he rebuilt his amplifier to go up to 11.

Dedication

Of course this book is dedicated to Katie, Erik, and Ally, the only energy sources that truly matter in my life. I would never choose any other alternative than what I have right now.

Acknowledgments

Many thanks to all those who have contributed to the material in this book, whether wittingly or not. Dick and Betty DeGunther; Professor Mitchell Weissbluth; Professor A.J. Fedro; John Lennon; Paul McCartney; Leland Stanford; Mike Pearcy; Jordan Cobb; Carl Marino; Eric Micko; Vikki Berenz; Connie Cowan; Betsy Sanders; Jim DeGunther; Sarah Nephew; Freddie Mercury; Dave and Perla DeGunther; Brad, Melinda, Samantha, and Emily Schauer; Chuck Albertson; Tilly and Evonne Baldwin; Joe and Marcia Schauer; and Kim and Gary Romano of Sierra Valley Farms. Thanks to Dr. Keith Kennedy and Watkins-Johnson Company for showing restraint above and beyond the call of duty. Thanks to John Steinbeck for making me understand what's important and what's not, and in the same vein, Derek Madsen.

Thanks to the excellent crew at Wiley: Mike Baker, Tracy Barr, and Jennifer Connolly. And thanks to my technical reviewer, Greg Raffio. Readers of this book will be amazed at how well it's written . . . it's not really my fault — I have the editors to thank for that. And thanks to Stephany Evans at FinePrint Agency for getting all the ducks in a row.

Thanks to all the *For Dummies* fans out there who have made the series what it is today.

Publisher's Acknowledgments

We're proud of this book; please send us your comments through our Dummies online registration form located at `http://dummies.custhelp.com`. For other comments, please contact our Customer Care Department within the U.S. at 877-762-2974, outside the U.S. at 317-572-3993, or fax 317-572-4002.

Some of the people who helped bring this book to market include the following:

Acquisitions, Editorial, and Media Development

Development Editor: Tracy Barr

Project Editor: Jennifer Connolly

Acquisitions Editor: Mike Baker

Copy Editor: Jennifer Connolly

Assistant Editor: Erin Calligan Mooney

Editorial Program Coordinator: Joe Niesen

Technical Editor: Gregory Raffio

Senior Editorial Manager: Jennifer Ehrlich

Editorial Supervisor: Carmen Krikorian

Editorial Assistants: Jennette ElNaggar and David Lutton

Cover Photos: © iStock

Cartoons: Rich Tennant (`www.the5thwave.com`)

Composition Services

Project Coordinator: Katie Key

Layout and Graphics: Samantha K. Allen, Reuben W. Davis, Sarah Philippart, Melissa K. Smith, Christin Swinford, Christine Williams

Proofreaders: Amanda Graham, Shannon Ramsey

Indexer: Anne Leach

Publishing and Editorial for Consumer Dummies

    Diane Graves Steele, Vice President and Publisher, Consumer Dummies

    Kristin Ferguson-Wagstaffe, Product Development Director, Consumer Dummies

    Ensley Eikenburg, Associate Publisher, Travel

    Kelly Regan, Editorial Director, Travel

Publishing for Technology Dummies

    Andy Cummings, Vice President and Publisher, Dummies Technology/General User

Composition Services

    Debbie Stailey, Director of Composition Services

Contents at a Glance

Table of Contents

Introduction

. . . .

*E*nergy is a big topic these days. Newspapers and magazines have articles about green living and energy efficiency in virtually every issue.
Everybody is concerned with the environment and the way that humans
are affecting the world that they live in. It seems a foregone conclusion that
the use of fossil fuels is causing global warming, or at the very least, making
global warming worse than it otherwise would be. The consequences of rampant fossil fuel consumption are not crystal clear, but the scenarios being
drawn by scientists point to a bleak future for humankind unless something
can be done to solve the problem.

Alternative energy encompasses the range of options that can be pursued to
make our world better. From solar to wind to alternative fuels for cars and
trucks, everyone can make a difference. The purpose of *Alternative Energy For
Dummies* is to explain the different alternatives and how they may be put into
play in the most effective manner. Whether you want to make small incre-
mental improvements, or grandiose, world-shaking changes, this book gives
you the info you need to understand what the pros and cons of each alternative option are.

Some alternative energy strategies can be implemented on a small scale by
individuals. You don't need to sacrifice your quality of life when you choose
to employ alternative energy schemes. Other alternatives, such as nuclear
energy, you won't be able to utilize in your own home. But you still need to
understand the pros and cons and be able to make intelligent, well informed
opinions about the alternatives that will very likely someday impact your
area.

About This Book

As you undoubtedly know already, many alternative energy books and
articles are on the market. So a lot of information is available. In this book, I
condense all the info about alternative energy into easily digestible chunks,
focusing on the those things that are most helpful in understanding alternatives to fossil fuels. In keeping with the *For Dummies* modus operandi, I give
you only the most relevant ideas, explain the advantages and disadvantages
of each, and offer the info you need to understand why and in what way certain schemes may be better than others.

There are a lot of energy-related books that get into far more detail than I do in this book. But this book is a great place to start, even if you're striving for more detailed information. I give you the big picture, and that's a great context to have before you dig in deeper. Although you won't be a certified energy expert when you finish this book, you'll be able to form logical, well-informed opinions about society's need to change our energy habits. You'll be able to discuss macro and micro energy issues and you'll understand life on the planet earth much better. That's a big claim to make, but energy is so entwined with life that understanding energy is a basic underpinning to understanding life. It may be said that life itself is energy . . . at least for some people it is — for others, a television is all that's needed.

In this book, you will find information like:

✔ What the term *energy* really means and how the basic laws of physics apply to all energy consuming processes, including alternatives. But don't panic: I leave out the complicated physics while giving you a good idea of what energy can and can't do.

✔ How the world came to the point that we are at now. Fossil fuels are, by and large, our main energy source, and there is a good reason for that. I also explain why it's going to be so difficult to change our economy over to alternative energy.

✔ Why fossil fuels are so dangerous for our society. Not only do fossil fuels cause untold and unnecessary pollution, but supply and demand problems also threaten the world's existing economy.

✔ How both traditional fossil fuel combustion processes work, as well as how the alternative energy options work. You'll also find info about the limitations and advantages of each.

✔ How you can conserve (which some would argue is an alternative in and of itself) and what the limitations of conservation and efficiency are.

✔ The range of alternative energy schemes currently being pursued and how society is grappling with the various problems inherent with the alternative energy schemes.

I wrote this book after having written two previous *For Dummies* books (*Solar Power Your Home For Dummies* [Wiley], and *Energy Efficient Homes For Dummies* [Wiley]). Both of these were very practical, with day-to-day tips and bulleted lists of information that allow a reader to grab their tool box and make effective changes in their homes. This book is more academic, and was much more fun to write. I expect that most readers will also have more fun reading it because knowledge and understanding are empowering. If I had my way, I would go back in time and finish my PhD in physics and be a college professor so that I would never have to do anything practical. I like knowledge for its own sake, and I feel empowered when I understand life in a particularly perspicuous way. My goal with this book is to empower the reader. When you're finished, you'll have a worldview that you did not expect, for there are some big surprises in these pages.

Conventions Used in This Book

For simplicity's sake, this book follows a few conventions:

- *Italicized* terms are immediately followed by definitions.
- **Bold** indicates the action parts in numbered steps. It also emphasizes the keywords in a bulleted list.
- Web site addresses show up in `monofont`.
- When this book was printed, some Web addresses may have needed to break across two lines of text. Rest assured that I haven't put in any extra characters (such as hyphens) to indicate the break. Just type in exactly what you see in this book, pretending as though the line break didn't exist.

What You're Not to Read

Although I'd like for you to find the topic so interesting that you wouldn't think of bypassing anything in this book, I realize that you may have other things requiring a bit of your time and attention. For that reason, I've made skippable information easy to identify. You don't have to read the following to understand energy alternatives. Although this information is interesting (if I do say so myself) and related to the topic at hand, it isn't vital, absolutely-must-know information:

- **Material in sidebars:** Sure, these are interesting. Some are fun. I like to think that all are helpful. But they contain info that you don't absolutely, positively, without-a-doubt need to know.
- **Paragraphs marked with the Technical Stuff icon:** Some people like details that only the technically minded or trivia-loving typically find interesting. If you're not one of these people, you can safely skip these paragraphs without missing any need-to-know info.

Foolish Assumptions

In writing this book, I made a few assumptions about you:

- ✔ You care about the world that you live in, and you care very much about how future generations will be able to enjoy their own lives on the planet earth.

- ✔ You understand that if humankind continues on its present course, some very big problems will only grow bigger and even more intractable and you believe that it's time to do something.

- ✔ Whether you believe in global warming or not, you do understand that humankind needs to change how we use energy. You don't have to be a tree-hugger or a "green" to see how the world is changing for the worse.

- ✔ You want to understand the relevant and important ideas about alternative energy as sensibly and efficiently as possible.

- ✔ You don't have an engineering degree, and you don't want to know every technical detail concerning the various technologies that I describe in this book. You simply want to understand an important subject.

Finally, because politics plays such a major role in the energy dialogue these days, let me explain where I come from, which may clear up some assumptions *you* have about *me*: I am a political centrist. I'm not a tree-hugger, nor am I a global-warming denier. I don't believe politics should play such a major role because I believe that humankind needs to change its ways. We use too much energy, and we use the wrong kinds. We put ourselves in jeopardy with our unending pursuit of fossil fuels. Yet while I do not advocate any political positions in this book, I take for granted the fact that changes to our energy consumption are necessary.

How This Book Is Organized

This book is divided into parts, each one dealing with a particular topic related to alternative energy. Each part contains chapters relating to the part topic. The following sections give you an overview of the content within each part.

Part 1: The Basic Facts of Energy Life

I begin with some historical perspective because that will give you an understanding of how we've gotten ourselves into the mess we're in. Humans have evolved in step with their energy consuming processes. When people first started burning fires, society evolved as health improved along with lifestyles. And by and large, the evolution of all human societies has depended on the effective use of energy, particularly in military ventures. In this part I also describe some very fundamental physical facts of energy. Most people have a very basic idea of what energy is, but in order to truly understand alternative-energy schemes, you need to understand how energy moves through a system. Energy can neither be created nor destroyed; it merely changes form. I describe precisely how this works.

Part 11: Digging Deeper into the Current State of Affairs

I describe existing fossil fuel machines and how they convert petrol potential energy into forms of energy that we use in our homes and to drive our vehicles. The picture of the world energy economy may seem bleak, but there are opportunities to introduce alternatives. I describe how energy efficiency and conservation can play an important role in reducing our reliance on fossil fuels. Finally, I survey the problems with fossil fuels, ranging from smog to global warming to simple problems of supply. Whether or not you believe that fossil fuels are poisoning the planet, the simple fact is that someday fossil fuels will run out.

Part 111: Alternatives — Buildings

In this part, which comprises a hefty percentage of the book, I describe the various alternative energy schemes that are being pursued and developed. I begin with nuclear power, which is going to play an increasingly important role in the future. Solar power is the "in" thing these days, and I explain the pros and cons and how you can use solar power in your own home (for more details, consult my book *Solar Power Your Home For Dummies* [Wiley]). Hydropower provides around 8 percent of American energy needs, but the future is very uncertain because of environmental questions. Wind power is a very good solution, but it's not for everybody. Geothermal energy is available in abundance in many parts of the country, and when it works it's a very good option. Biomass is an interesting alternative because it's so widely available and makes good use of materials that have been, heretofore, simply thrown away. Wood burning, when done properly, can be an excellent alternative for those who have a ready supply, but there are some major problems that need to be understood in order to do it right. And finally, I touch briefly on fuel cells, which hold the most promise of any of the alternative-energy solutions.

Part IV: Alternatives — Transportation

The majority of fossil fuel consumption occurs in the transport sector. Cars and trucks spew billions of pounds of emissions into the atmosphere every year. I describe some alternative fuels that are increasingly being used, such as corn ethanol and biodiesel. I describe how these fuels are best used, and when and where. I describe how hybrid autos work, and how all-electric vehicles and fuel cell–powered vehicles work. I give you some guidelines to use if you're interested in investing in alternative vehicles. And finally I survey some of the more exotic alternative transportation systems being developed.

Part V: The Part of Tens

Like every *For Dummies* book, this part includes quick resources that provide plenty of information and sage advice compacted into few words. Want a list of the best ways to invest in alternative energy? Or maybe you want to understand some of the more prevalent myths about alternative energy. Perhaps you want to help to change society? I give you the nitty-gritty, in as few words as possible.

Icons Used in This Book

This book uses several icons that make it easy for you to identify particular types of information:

The Technical Stuff icon lets you know that some particularly nerdy information is coming up so that you can skip it if you want. (On the other hand, you may want to read it, and you don't actually have to be a nerd. You only have to be able to read.)

This icon indicates a nifty little shortcut or timesaver.

Look out! Quicksand is afoot. You don't want to skip over the warnings. They point out dangers to your health and well-being, your property, or your bank account.

 This icon highlights important information to store in your brain for quick recall at a later time.

Where to Go from Here

Some of you may look at the table of contents and skip straight to a particular section that addresses a topic you wish to understand without resorting to the fundamental background information. Others of you may start in Chapter 1 and work your way diligently through the book all the way through Chapter 23 Either approach is just fine and dandy because all *For Dummies* books are structured so that you can jump in and out or read them straight through to get the information you need.

 The goal of this book is to impart a very basic and broad-ranging understanding of a major problem facing humanity. In this regard, it's probably better to read the book from beginning to end because there is coherency and a logical flow to the arguments being presented. Understanding how current fossil fuel technologies work may not be necessary to understanding alternative energy, but the word "alternative" does have a particular meaning that is directly related to fossil fuels. Alternative energy schemes are simply alternatives to the fossil fuel paradigm that the world operates under.

Part I
The Basic Facts of Energy Life

"I don't know much about alternative energy sources, but I'll bet there's enough solar power being collected on those beach blankets to run my workshop for a month."

In this part . . .

Alternative energy is defined differently by different people. Obviously, the definition you get depends on the context and who is doing the defining. In this book, I define alternative energy as any energy source that is an alternative to the fossil fuels that rule the world today. In Chapter 1, I describe how fossil fuels have evolved to the point where they are the overwhelmingly dominant energy source. In Chapter 2, I explain what the term "energy" actually means — info you need to know to understand how the different alternatives compare to the current system that relies on fossil fuels. In Chapter 3, I get into some hard facts of energy consumption, important background information for understanding how energy moves through a system and what happens to energy once it's "used."

Chapter 1

What a Mess!

For the most part, producing energy and consuming energy is a very dirty business. Although you all may have a sense of this, the extent of the problem seems to be a political question open to debate. What should be done about energy consumption? Use less? Use different resources? How can new energy sources be best invested in? And perhaps of most importance, what types of new energy sources should be invested in? And what about global warming? Can anything really be done about it? What will new energy sources and the combating of global warming cost both the society as well as individuals?

A lot of solutions are being tossed around, but it's nearly impossible to separate the wheat from the chaff. And to make matters worse, political fringes screech from the sidelines in preachy tones that tend to turn people off to the point where they just plain don't want to listen any more. This is a system guaranteed to create gridlock and partisanship, and that's been the norm for so long that everyone is just plain used to it.

Leaving the important questions up to the politicians hasn't worked too well, and so it's incumbent on everybody to understand the issues so that informed decisions can be made when voting for candidates with varying views concerning both the problem and the solutions. Informed decisions also help you to decide, on a micro basis, what's best for your homes and your families. This chapter gives an overview of energy use — as well as the system's problems — from the past and present as well as what can be done in the future.

Understanding Where Society Is and How It Got Here

The fact is, energy is a critical component of your lives and your economy as a whole. You use energy in virtually every endeavor you engage in, whether you're aware of it or not. Life expectancy in the U.S. has increased 66 percent over the last century (from 47 years in 1900 to 78 years today). Americans are living longer, healthier lives, and for the most part Americans are more secure and knowledgeable about their world. Life is just plain better than it used to be, and this has been made possible by advances in medicine and technology — advances that all took a great deal of energy.

To make any kind of claim that energy consumption has been anything but advantageous to humanity completely misses the point. The problem is not with energy, it's with the way energy is used and which types are used.

It's only becoming evident now that energy use is a zero-sum game: You pay every bit as much as you gain, but the terms under which you pay are still not clear. Of course, you pay for each gallon of gas you use, but we're also learning that we are paying in environmental costs and health costs. The most fundamental concept that this book can teach is this: The U.S. (and by extension, the world) does not have an energy crisis. Rather, it has an *environmental energy policy crisis.* The U.S. has as much energy as it wants to use. The question is how to use it and what kinds of limitations should be set in terms of how the environment is affected through that energy use. To answer that question, you need to get a handle on energy use in the past, present, and future.

Historical trends of energy use

Humans have evolved in step with the sophistication of their energy consumption. Human populations, quality of life, and life expectancy have increased as energy sources have become more sophisticated.

Think about it: Early man couldn't even light a fire. Many froze in the winter, with only wooly mammoth skins to keep them warm, and the quality of life was not much different than that of animals. Once fire was discovered and humans were able to create flames at will, humanity began a gradual but consistent climb from savagery to what it is today. Upon the advent of fire, humans could warm themselves and cook their food. This began the consistent push toward bigger and better cultural and material gains, and it lead to healthier, happier lives.

Throughout time, the human population remained steady for the first 1,500 years and then began a steep, consistent climb. The increase is due largely to the availability of versatile, convenient energy. As controlled, or useable, energy became more prevalent, the population expansion accelerated.

Table 1-1 shows how population and energy consumption per capita have increased through the ages. Continue this trend out a hundred years, and it suggests that the only thing humans will be doing a century from now is consuming energy, 24/7. Regardless whether you're willing to take that leap intellectually, the fact is, humans use more and more energy every year. This growth can't be supported, unless society comes up with alternative sources and consumption habits; otherwise, fossil fuel reserves will be depleted by the year 2050.

Table 1-1	Population and Energy Consumption over Time	
Date	*Population (in billions)*	*Consumption in kWh/day*
5,000 bc	0.1	9.4
0 ad	0.3	10.1
1850	1.3	12.0
1980	4.4	51.0
2000	6.0	230
2050	9	1,000?

You can see how greater populations results in higher energy consumption. You can also get a feel for how daunting and necessary it is to find a workable solution for energy consumption.

Energy use today

Today, Americans consume around 100 Quads (quadrillion Btus, or British thermal units; head to Chapter 3 for information about energy measurements) of energy per year. (This number includes only sources of thermal energy, such as gasoline, natural gas, coal, and so on, and not the vast carbohydrate network that supplies our food chain.) So is that a lot? Here's 100 Quads worth of energy in units that resonate: Fifteen large horses labor, day and night, 24/7, for each U.S. citizen.

The breakdown of U.S. energy consumption is roughly 40 percent electric, 30 percent transport, and 30 percent for heating. While we burn most of this energy, only about 30 percent goes to direct heating. The rest is used to turn shafts to make electricity, and to turn shafts to move our cars. Eventually, all of this energy makes it back into the environment in the form of heat. In fact, there is far more heat pollution from energy consumption than any of the chemical types of pollution emitted, but Mother Nature can absorb all of this heat without the problems that chemicals create.

Figure 1-1 shows how primary energy sources are used in the American economy. As you can see, coal, oil, gas, and shale (the fossil fuels represented in Figure 1-1) are the most commonly used energy sources today and the root of many of our current problems. Mineral fuels like uranium are used for nuclear fission and fusion.

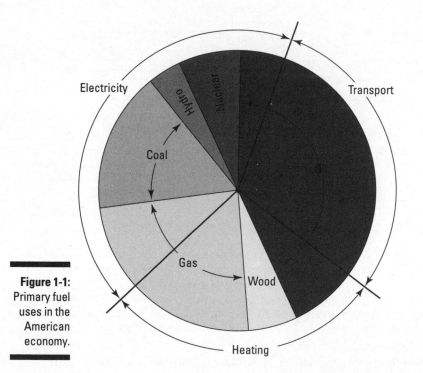

Figure 1-1:
Primary fuel uses in the American economy.

For more information on the current state of affairs of our energy use, head to Part II.

Looking ahead to more energy use and waste

As the human population grows, the total consumption of energy grows even faster. Bottom line? We humans seek comfort and consistency in our lives. Energy provides this, and so humans seek energy. The reason so much more energy is being consumed per capita is that ordered energy (go to Chapter 2 for an explanation of this concept) takes a lot of raw energy to create. Raw forms of energy, like firewood, give way to more sophisticated forms, like electricity. That's evolution, in a nutshell.

Humanity strives for consistency. Consistency is order, and that's what energy gives us. If we want more consistency, it will take more energy. And more and more all the time. Human history suggests that our energy consumption is only going to increase, and when impediments to this growth occur, problems follow.

The following sections delve into some of those problems.

The cost of energy producing processes versus raw energy

As more sophisticated energy-consuming processes are developed, the cost of those processes has less and less to do with the cost of the raw energy required and more and more to do with the equipment that creates the finished energy product. You have to consider the total effect of the energy-consuming process. The order achieved from an energy-consuming process is important, not the actual raw energy that is being consumed.

Invested energy is the energy that's used to manufacture a product (in the literature, you may see this referred to as *embodied energy,* or grey energy, but it all means essentially the same thing). Most people don't consider invested energy when they're analyzing energy-consuming processes; they only consider the raw fuels that are used by the machines that use energy. But a complete energy picture requires consideration of invested energy.

Decreasing costs, increasing demand

When more efficient machines are devised, the cost of operating those machines decreases, and demand goes up. This is simple economics referred to as *cost elasticity:* The cheaper a commodity becomes, the more it is consumed. For example, more fuel-efficient cars result not in less consumption, but more consumption, because people can afford to drive their cars more.

The inevitability of waste

Waste is a necessary part of every single energy process. It's a simple fact of physics. Therefore, the pursuit of eliminating waste is fruitless, and misses the point. In fact, the more ordered energy becomes, the more waste that is inherent in the process of creating that order. Chapter 3 has more on this topic.

Putting Society Over a (Oil) Barrel: Problems in the Current System

Over the last century, humankind experienced an unprecedented expansion of industrialization. Here's a perspective: Only 100 years have passed since autos have even existed, and electrical power didn't exist in a majority of the country until 50 years ago. Now consider:

- ✔ Industrialization is a global phenomenon so energy resources are being consumed at an unprecedented rate the world over.

- ✔ Japan is the world's second largest economy, and China is growing into a tremendous world power, consuming resources faster than any country the world has ever seen.

- ✔ U.S. consumption of petroleum has risen steadily since 1950, and promises to continue unabated.

- ✔ In 2000, the U.S. consumed 20 million barrels of crude oil per day. That's a lot of crude.

- ✔ U.S. net fossil fuel imports fell during the '70s as a result of the OPEC oil embargo, but have risen sharply since the mid 1980's. In 2005, over half of the crude oil consumed in the U.S. was imported, and it's projected that this number will increase to over 65 percent within 20 years.

Overreliance on fossil fuels

All of the growth and expansion has been shouldered by fossil fuel energy sources, which creates all sorts of havoc: Politically it makes us beholden to foreign nations. Economically it creates uncertainty in markets, and wild price fluctuations result. Economies need consistency in order to thrive, and there is no telling how much the fossil fuel price fluctuations have cost the U.S. economy in terms of lost growth potential.

Declining supplies

While consumption has skyrocketed, U.S. domestic supplies of crude oil have declined in the last half of the century, and after checking out the following facts, you can see that at some point U.S. domestic supply will simply run dry:

✔ **At present, U.S. production of crude is only 1.9 million barrels a day.** Current estimates put U.S. domestic reserves of crude petroleum at 21.3 billion barrels. But current production rates are around 8 percent of the reserves per year. This means the U.S. will run completely out of domestic crude in about 12 years

✔ **Current estimates of U.S. natural gas reserves are 192.5 trillion cubic feet.** But production from these reserves is growing at a rate four times that of new discoveries so the situation is only growing worse. At present, production represents around 4 percent of proven reserves per year, which suggests that the U.S. will run out of domestic supplies in 25 years.

Even if ANWR is opened up to drilling; even if offshore drilling platforms are allowed to proliferate around the coastlines; even if new domestic reserve discoveries exceed the wildest expectations; and even if future generations of extraction and processing equipment exceed expectations, the U.S. will run completely out of domestic supplies for its most basic forms of raw energy.

The U.S. has two options in the face of declining reserves:

✔ Seize foreign reserves, by war if necessary. Although this may sound outlandish, when the U.S. economy starts to grind to a halt, things will get truly serious.

✔ Develop alternative energy sources.

Getting better at finding and using fossil fuels

The supply of energy is not determined by how much is available, or what's out there, but by humankind's ingenuity in getting to it, and using it. Semiconductors and computers have made oil-drilling machines infinitely more intelligent and effective than they were 100 years ago. And semiconductors and computers now account for nearly 15 percent of all the electrical consumption in the U.S. Energy begets energy. In short, we have grown much more efficient at extracting raw energy.

A word or two about global warming

I don't take a definitive position on global warming one way or another in this book. My opinion is that it's not really relevant in encouraging people to conserve and practice efficiency. The fact is, the world uses too many resources, and in the process ends up altering the planet in material ways that will affect future generations in one manner or another. Global warming is just one way that humans might be affecting the planet, and in my view the overwhelming attention placed on this one aspect of our environmental defamation does a disservice to the other aspects, and the overall balance and harmony that mankind should be striving for. What if it turns out that there is no global warming? Should society then go right back to the old days of unlimited exploitation of resources? Of course not.

I also find it frustrating that most people seem to believe wholeheartedly that global warming either does or does not exist. There seems to be no middle ground. The fact is, the data, while compelling, is still inconclusive, and both sides have valid arguments. The earth has always warmed and cooled. There have been over 600 warming and cooling cycles, and none of these have been attributable to manmade pollutants. It is theorized by some that the dinosaurs' demise was brought on by a meteor that hit the planet earth, causing tremendous densities of airborne particulates that changed the environmental ambient in catastrophic ways. Warm-blooded creatures thrived in the new world order, and this brought on mankind's ascendance. So within this theory, humankind is a product of a natural global warming event. It would be ironic if mankind ends up ensuring its own demise by a manmade global warming event.

Here's an interesting question for those who believe that mankind should strive to neutralize its effects on the planet. Suppose NASA determined that a large meteor was going to strike the earth. And further, it would be possible to launch a rocket with a massive nuclear bomb that could destroy the meteor. Should we do it, or should we simply let nature take its course?

In conjunction with this trend, technology has provided the means to refine energy to the point where it can be controlled for even the most precise uses, like resurfacing the human eye so to see without eyeglasses, or creating microprocessors so precise and controlled that even microscopic variations in the semiconductor substrates are smoothed over, resulting in ever faster and better computers, that in turn, make it possible to harvest even more energy, and with more efficiency.

Imagine if all of this ingenuity and inventive spirit were channeled into the pursuit of alternative-energy sources. Most of the advances in technology come from the U.S., and if the country were to focus on developing alternative-energy sources, a huge boost to the economy would result. The U.S. could lead the world in exporting alternative-energy know-how, and the equipment to achieve the desired end results.

Rising to the Challenge: Balancing Fossil Fuel Use with Appropriate Alternatives

You solve energy problems by expending even more energy, not less. You need to devise alternatives that offset fossil fuel addictions, and invention and development take a lot of energy. Infrastructure takes energy. All human advancement requires energy in increasing amounts. Because the only consensus seems to be that fossil fuels are not the answer, you have to wonder what form this energy will take in the future. The following sections provide an overview of the alternatives as well as factors to consider when choosing alternative sources of energy.

This book is all about alternatives to the status quo. Alternatives are not the end all; they will never displace fossil fuels. The solution lies in a combination of doing better with fossil fuel use and developing alternatives when that's appropriate. In Parts III and IV, I describe the alternative technologies in detail, explaining when these technologies are useful and when they're not. Alternatives will require sacrifices not only in terms of monetary costs but also in terms of changing lifestyles.

Looking at the local impact

In addition to the pollution mitigations and political desirability of alternative energy sources, there are attractive local impacts:

- ✔ Local jobs
- ✔ Sustainable economy
- ✔ More money stays local, instead of moving to the Mideast
- ✔ Less air pollution, lower health burdens
- ✔ Diversification of the energy-supply options
- ✔ Security to the U.S. economy — the economy is more controllable if it doesn't rely on foreign countries for energy supplies.
- ✔ Increasing supply of energy options reduces costs by increasing competition, making inexpensive energy more widely available

Economically, none of the alternatives can compete price-wise with fossil fuels, but that all depends on one's accounting system, for fossil fuels are subsidized in many ways by the governments of the world.

Adding up the alternatives

As stated earlier, energy is not running out, nor will it ever run out. The problem is not that less energy resources are available, but that the political and environmental consequences of the current energy consumption, well, stink. Hence, the drive for alternatives. The following sections introduce the alternative energy candidates. You can find out more about these options in Parts III and IV.

Solar power

Solar power, discussed in detail in Chapter 9, uses sunshine to create both heat and electricity, as well as passive heating and cooling effects in buildings. Although there are other ways to take advantage of solar power (think photosynthesis, for example), the one I focus on in this book is the direct conversion of radiation. This includes photovoltaic panels and solar liquid heating schemes. Large scale solar farms can provide entire communities with enough electrical and heating power to make the communities self-sufficient.

Nuclear power

Nuclear power harnesses the tremendous energies from both the splitting and fusing of atoms. In some books, nuclear is not considered an alternative energy source, but my interpretation is that alternatives are those that do not emit the fossil fuel pollutants that are causing so much environmental harm. So I include nuclear energy in the alternative energy pantheon. Find out more about nuclear power in Chapter 8.

Solar's role in other energy sources

Solar power is a key component in other energy sources:

✔ **Photosynthesis (a plant's ability to convert sunlight into useable energy):** Plants grow and may be combusted as biomass (like ethanol, or wood). Animals eat plants, humans eat animals and plants. There is also energy available from fermentation and anaerobic decay of biomass.

✔ **Oceanic.** Waves may be harnessed for energy production. Currents are capable of driving hydro turbines. The thermal differences between different regions may be tapped with heat exchange mechanisms.

✔ **Hydropower.** Solar radiation evaporates water, which becomes rain, which becomes rivers and streams that can be dammed up and outfitted with turbines and generators. See the section "Wind and hydropower" and Chapter 10 for more on this energy source.

Wind and hydropower

Wind power derives from windmills placed in locations with a lot of wind. Luckily for the U.S., there are plenty of suitable sites.

Hydropower comes from dams which provide high pressure water flows that spin turbines, thereby creating electricity. It can be exploited on both a macro level (huge dams can be built to create statewide electrical power on America's biggest rivers) and on the micro level (people can put hydropower generators in backyard rivers and streams). For more info on both of these energy sources, head to Chapters 10 and 11.

Geothermal

Geothermal power, the topic of Chapter 12, takes heat from the earth and redistributes it into a building, or uses the heat to generate electrical power. It's available in tremendous quantities, but it's difficult to extract and takes a lot of capital equipment. On a more general level, heat pumps (the kind in many homes) are a source of geothermal energy, so geothermal energy can be practical and effective on a micro level.

Biomass and wood

Biomass is sawgrass, mulch, corn, and so on. These materials are either burned in their raw form, or processed into liquid fuels or solid fuels. Wood, the most common biomass, is used to heat homes throughout the country. (**Note:** Some books distinguish wood from biomass, but I don't make that fine a distinction.) Chapters 13 and 14 explore these topics in more detail.

Hydrogen fuel cells

Hydrogen fuel cells, in a nutshell, produce electrical power from nothing more than hydrogen, which is completely free of carbon. The exhaust is water, and what can be more natural than that?

Hydrogen fuel cells combine oxygen and hydrogen to produce water and electrical energy. Sounds simple, and there's an amazing potential to solve a lot of the world's environmental problems, should fuel cells pan out like some people think they will. The technologies are years off, however. And there are some major difficulties that may never be overcome. But the promise remains bright, and a lot of development money is now being invested in fuel cells. Go to Chapter 15 for more information.

Electric vehicles

Electric vehicles use only electricity to power the drive train. The electricity comes from batteries, which are heavy and cumbersome, but battery technologies are getting better and all-electric vehicles are now becoming economically competitive with conventional, internal-combustion vehicles. It should be mentioned that electric vehicles need to get their electrical power from somewhere, and that "somewhere" is likely the power grid, which itself consumes a lot of coal, and emits a lot of pollution. Chapter 18 has more info.

Hybrid vehicles

Hybrid vehicles are a combination of electric and internal-combustion powertrains. When power requirements are low, the vehicle operates in electrical mode. When more power is needed, or when the electrical batteries are near depletion, an internal-combustion engine provides power. Hybrids, discussed in Chapter 19, offer much higher MPG ratings than conventional transportation.

Biofuels

Biofuels, discussed in Chapter 17, are made of biomass products such as corn. Corn ethanol is now being added to most gasoline supplies in the United States. Despite the high energy consumption in the refining process, biofuels allow the U.S. to import less foreign oil, and so the political effects are desirable. Biofuels may either be used in their pure form or mixed with fossil fuels.

Evaluating the alternatives

There is no such thing as a free lunch. Every energy source has pros and cons, and trying to decide how best to provide the power an economy needs is a complex problem. Many believe that the current energy predicament will be solved by weaning society away from petroleum consumption, but even as people develop new alternative sources, the problems don't go away; they simply change in nature.

In evaluating alternative energy sources, here are some important factors to consider.

Combustion versus noncombustion

The majority of our energy sources produce power through combustion processes (burning) that require a burn chamber, oxygen, and exhaustion capacities. From time immemorial, humans have burned wood for fires, and the process was simple: Pile some wood, light it on fire, hang around nearby. Modern combustion processes are engineered to be more efficient (modern wood-burning stoves are around 100 times more efficient than open fires, for example), but the combustion processes, regardless of how efficient they are, are notorious pollution sources.

Noncombustion processes, such as solar power and nuclear, don't exhaust pollutants the same way that combustion processes do, but they entail their own problems. For instance, solar photovoltaic (PV) panels require a lot of energy to manufacture, and most of this energy comes from electrical power which mostly comes from coal combustion. So while solar is pollution free in its on-site implementation, it entails a lot of pollution in its manufacture. Other noncombustion energy sources such as wind and hydropower also require a great deal of energy to manufacture the capital equipment needed to make things work.

Raw material issues

Every energy production plant, whether solar or a woodstove, needs raw materials. In the case of solar, the raw materials are free. In the case of a nuclear power plant, the raw materials are uranium rods, which must be meticulously refined and manufactured. In fact, the total cost of an energy process has less and less to do with the raw fuels. Capital equipment is expensive, and is usually the most influential component in a cost/analysis equation.

The degree of refinement of the energy

Woodstoves provide heat, and in a rather coarse fashion. Solar PV provides high-grade electrical energy. Wind power also provides high-grade electrical energy. In the case of the woodstove, the heat is the desired end product, and heat is very coarse yet effective. Electrical energy is very refined and convenient. Every energy-consuming process requires a certain degree of refinement of the energy, and the refinement itself takes energy. If it's possible to adopt policies that promote less refined energy, everyone is better off.

The level of current technology

It takes time for new technologies to reach the market, and it takes even more time for wide-ranging acceptance and use of a new technology. People don't simply discard their current systems because a radical new technology is developed. They wait until their current equipment breaks down, or is no longer economical to use before they invest in new systems.

Pollution and environmental impacts

Every energy-producing and -consuming process leaves a residue of some kind on our planet. Alternative-energy schemes are not all pollution mitigation marvels. Wood stoves, for instance, can be one of the most polluting energy sources if the wood is burned inefficiently. And different types of pollution cause different types of problems.

Economics

Let's face it; most people are concerned exclusively with economics and are only interested in seeing their net costs decrease for energy consumption. Because of this, the government steps in with taxes, rebates, and other forms of subsidies in order to achieve in the market what they deem desirable, namely lower pollution levels and freedom from foreign oil. If the government didn't mandate economic changes to the playing field, fossil fuels would never yield to alternatives.

Politics

And of course, since government is going to lead the way into the alternative energy future, politics plays a very large role in which alternative solutions get the most play. Green politics is becoming an increasingly powerful and influential part of every government operation.

Chapter 2

What Energy Is and How It's Used (Politically and Practically)

In This Chapter

▶ Wrapping your brain around energy as a fundamental life force

▶ Understanding how the refinement of energy impacts energy consumption

▶ Peeking at political positions on fossil fuels and the potential of alternatives

*T*here are a lot of myths concerning energy. In this chapter I describe exactly what energy is (and isn't), and why it's so important in every aspect of life. Prior to understanding the nitty-gritties of energy, it's first necessary to gain a good understanding of the abstractions of energy.

In this chapter, the basic concept of energy and power evolves into a more focused understanding of useable, or *ordered,* energy. What humans seek, as they use energy to make their lives better, is not just energy, but refined and controlled energy.

Defining Energy: Grab Your Hats, Folks

Most people think they have a good idea what the term *energy* means: the gas you put into your car, or the power that comes into your home via electric wires that makes your television spew out such meaningless drivel. For the most part, your understanding of energy's *functional* nature is correct. But the term is actually much broader than that. In fact, at the highest level, energy is the essence of life in the universe, and ordered energy is the key to human advancement.

The following sections break down the concepts of how energy is behind life and all that you do. These sections show you how simple and complex energy is, which explains why the entire concept of energy can make your head swim. No worries though — although I dive right in, I give you plenty of water wings and inner tubes to keep you floating right along.

Life itself is energy

You probably have an intuitive idea of what energy does, namely perform work, or make things work. It takes energy to make something happen. And it takes energy to make something stop. In short, almost every single life process takes energy.

Even processes without life take energy. Imagine a planet without life. Only physics — natural laws that govern the universe — survives (and then, of course you could argue that if nobody's around to note these physical laws, then maybe even physics doesn't survive without energy either . . . is your head spinning yet?). In physics courses, almost every single process and equation is described in terms of energy. The way energy moves through a physical system defines how that system works, in minute detail. When you see a picture of a physics professor standing in front of a chalkboard with a whole bunch of abstract mathematics, for the most part the math is describing how energy proceeds through a physical interaction and how the different parts of a physical system are related in terms of energy.

The following list explains how it may even be said that life itself, from beginning to end, is a form of energy:

- **As life evolved on our planet, more and more complex life forms became possible.** What has made more complex forms possible is the information storing mechanisms hidden within DNA structures, which themselves are essentially energy centers. DNA is a sugar, and sugar is energy. The passing of the information contained in DNA is inherently linked to an energy source that makes life possible. In this vein, information itself is a form of energy. At the very least, there would be no transfer of information without energy. There is no transmission of computer data without energy. In fact, there is nothing without energy.

- **Complex life forms are always characterized by complex energy schemes.** Warm-blooded creatures are a good example. The evolution of warm-bloodedness enabled these animals to overcome cold weather and move about more freely. But there was a price to pay as well. Warm-blooded creatures require much more energy to live and thrive; they must eat a lot more than cold-blooded creatures in order to maintain. The complexities increase dramatically, and the energy systems required to support this complexity also become more complex.

- **Life forms consume other life forms; a form of energy mutation.** The major preoccupation with all of life is the attainment of more and more energy. It's impossible to spend a single moment of life without consuming energy in some form or another. The process by which energy is extracted when a life form eats another is not much different than the process that automobile engines use to extract energy from a gallon of gas. The chemical bonds of the raw fuel are broken down and in the process energy is released. In fact, that gallon of gasoline was, very long ago, a plethora of life forms which died and decayed and turned into crude oil deposits beneath the surface of the earth. Humans eat life forms for sustenance and burn life forms for heat and utility. If the world were truly fair, other life forms would eat humans, although the flavor would probably not be so great, with all the crud we put into ourselves these days. Crud, by the way, which is made possible by energy-consuming inventions that we hold so dear.

✔ **Perhaps the human soul is nothing more than an ineffable form of energy.** (I'm not getting religious here — I'm simply making an observation). Energy can neither be created nor destroyed but simply changes form (this is the essence of the first law of thermodynamics; see Chapter 3) — a fact that can be observed in the physical world. If the life-force of a living creature is an energetic order, then we can say that death is not the end all, that the energy that animates the living creature — what many call the soul — continues on.

The fact is, energy production and consumption is critical to all life on the planet earth, but it's particularly critical to human life. The reason humankind has thrived and dominated the planet is due to our ability to devise clever energy exploitation mechanisms. The goal? To create order out of chaos (see the section, "Energy is order," below).

Energy is order

It takes energy to make the world conform to mankind's wants and needs, and that's the bottom line. By controlling energy, humans have been able to alter their universe, and ultimately that seems to be what humanity is all about: controlling life with energy and power. Check out these examples:

✔ Humans can obtain what they want when they expend energy.

✔ Humans can protect themselves from their enemies when they harness energy into formidable weapons.

✔ Humans assure a consistent supply of comfort and health by expending energy.

In this book, what I'm really concerned with is not energy, but *order.* Looking at energy as a barren physics concept only leads to confusion because citing energy statistics tells you only part of the story. Sure, you can measure energy using tactile devices and scientific terminology (which I get into in Chapter 3), but what you should ultimately be interested in is not how many joules you've burned or how many kilowatt-hours a process consumes. The notion of efficiency simply means that mankind accomplishes its goals using less energy.

To give you a better idea of what I mean by ordered energy, I describe some processes in detail in the following list. Think of the details not in terms of a lecture on how the process works but in terms of how so many ordered steps go together to output specific uses of energy . . . you'll never turn on your vacuum or wash your clothes with the same mindset again! To sharpen the concept of ordered energy, consider how energy consumption creates order:

- Your washers and dryers clean your clothes, but you don't really need to clean your clothes nearly as often as you actually do. What you seek is cleanliness, and perhaps you wish to smell better in the process as well. You don't want wrinkled shirts, nor spotty pants. Dirt gets onto you in all walks of life, and that is disorder. What you want is to reverse that disorder. You want to control the environment and it hardly matters if you need to or not. You simply want to, and that's enough rationale.

- You use vacuum cleaners in your homes, and you wash your cars using water pressure and soap which came from a factory that used energy in the manufacturing process. The soap was delivered to the store via a truck, which consumes energy. Energy is consumed in every process that delivers the soap to the store, and then more energy is consumed in bringing that energy home and making good use of it. Even after you're finished, energy is expended in the disposal of the soap.

- You use electricity for your televisions, which take complex signals received from a variety of noise-embedded sources and turn them into ordered images that you watch, and listen to. Think of all the disorder associated with a television signal, until that last instant when it's all turned into a coherent, ordered image. A great deal of unseen energy is expended just so you can enjoy that final burst of coherence — or, if you watch network television on any given night, incoherence.

- You use lasers to change the curvature of your eyes, so that you can get rid of those pesky glasses and contact lenses. The laser process begins with a big fire in a utility company furnace somewhere far away. The fire boils water, which forces a big turbine to spin, which in turn generates a coarse form of electricity. That electrical power is fed down a long stretch of high voltage power wire, into a local substation where the characteristics of the power are altered. Finally the power enters your optometrist's office, but that's not the end of the story. The power is filtered through a series of high-precision electrical networks until it finally reaches the laser substrate, which in turn creates highly ordered photons that interact with your eye in such a way as to reshape your optical lens. Photons are energy, and they are very precisely directed toward burning a small portion of your eye, so as to reshape it. At each stage of the energy process, there is more order (and a certain degree of waste, since nothing is efficient). The entire process culminates in a very focused burst of energy, and you can see without eyeglasses once again.

✔ You build homes to shelter against the random elements. Homes and buildings take a considerable amount of energy not only to build, but to run. You have to maintain the home against assaults from Mother Nature's energetic furies.

✔ Transportation gets you to and fro, in an orderly, safe, fast, and efficient manner (okay, so maybe I'm not including public transportation here, but just think of the energy you expend when your blood begins to boil at the fact the train is late!). When you can come and go as you please, it's a form of order. It takes a great deal of energy to afford society this luxury.

✔ Computers have become geometrically faster, with a commensurate increase in the power they consume. The Internet now consumes around 8 percent of the electricity produced, and computers in general consume around 15 percent. From all of this comes order:

- Work is more productive and faster, not to mention more enjoyable and rewarding.

- The American economy produces more economic output per unit of energy than any other economy the world has ever seen.

- And the demand for energy to run information systems is increasing faster than any other segment of the economy.

Americans may be energy pigs, but Americans also produce quite a bit with their energy.

From simple to complex: The evolution of energy

Energy may be categorized as simple or complex, low-grade and high-grade. The simplest energy process that humans exploit are wood fires. Stack some dry logs, light them on fire, sit back and enjoy the heat. Wood burning fires are an example of low-grade energy, because the benefits are very coarse, and the amount of useable work performed is limited to very unfocussed heating. Most of the heat generated in a wood fire is lost to the environment, as opposed to directed onto a human's skin. Back in the day, if more heat was needed, you stacked more wood, caring nothing about how efficient the process was. That resource was seemingly infinite and was treated as such throughout most of human history.

As energy processes evolved and became more directed toward a specific end, human civilization evolved as well. Humans became better at burning fires, not only extracting more heat from smaller stockpiles of raw materials, but also figuring out how to insulate and keep that heat around longer. Then humans figured out how to build steam engines, with their logical valves and timing mechanisms, and this launched the industrial age and made life more pleasant and healthy. Then along came the internal-combustion engine, then electrical generation, then radio, radar, television, lasers, and so on.

With each successive increment of energy refinement, human life evolved and more comfort and convenience resulted. Now that the energy situation is changing, society struggles to come to terms with the limitations that are being imposed. Society is struggling with the notion that energy supplies (insofar as the type of energy that is used) are limited, and that the energy being used is affecting the environment in negative ways. Basic attitudes about energy are changing in dramatic ways.

Trying to Tap into and Refine an Infinite Energy Supply

There is no shortage of energy, and there never has been nor will there ever be. Over 35,000 more energy reaches the earth's surface from the sun than humans use in all their aggregate processes. That's clean energy, and completely natural. Geothermal energy comes from heat within the earth, and there is an abundance of that particular energy source. (Ultimately, all energy sources derive from the sun. Fossil fuels were once life forms which relied on the sun for warmth and energy. Plants rely on the sun for their basic energy. Animals eat plants, and so on. And there is no shortage of sunshine.) So, although infinite energy supplies abound, tapping into and refining those supplies is the issue, as I explain in the following sections.

Refining ordered energy

So while there's no shortage of energy itself, there is a shortage of useable forms of energy. The latest projections indicate that there is enough fossil fuel to power the world economy for at least another 50 years, and this is fuel that comes from proven reserves. With each passing year, more reserves are found, and with vast improvements in energy technologies, these reserves are cheaper and cheaper to extract. What's interesting is that it takes so much more energy to extract previously inaccessible reserves. The more energy that's found, the more energy that's used, and energy consumption continues to grow with the invention of bigger and better machines for energy extraction, and these machines consume huge amounts of energy in the process (refer to Chapter 1).

The problem for humanity, then, is not energy, per se, but *ordered energy*. We need more refined energy, in the form of electricity, and high-quality fuels to burn in our autos and jets.

The advances of the industrial revolution were obtained not so much by the inventions (although those played an important part) as by the ability to create more ordered and refined energy sources.

Jet engines, which are capable of huge power outputs within small sizes, take ordered energy a big step further. They are much more efficient at extracting energy from a pound of raw fuel. The reason jets can travel so far and fast is not energy, per se, but ordered energy. This is very important because when a jet plane first takes off, it must carry all of its own fuel. Jet engines work so well now because the fuels have been refined and improved, and it's possible to extract much more energy per unit weight from these fuels. By ordering the energy producing process much better, bigger jets are possible, as well as more efficient air travel, which allows for cheaper tickets and happier customers.

There is no shortage of energy; there is only a shortage of ordered, useable energy, and in providing ordered energy there is inherent waste. For example, it takes around two gallons of gasoline to get one gallon to the fuel pump near your house. The solution to the energy problem is finding better ways to order the energy sources that society has access to.

Bridging the gap between policy and pollution

The U.S. does not have an energy crisis. It has an environmental energy policy crisis. To make my point, I go back to solar. There is no shortage of solar power, but the vast majority of it is simply an incoherent barrage of unordered photons. You can lay in the sun, and warm yourself, but you can only do this on a warm, sunny day. In order to harness solar power, you need intelligent machines that can convert the incoherent barrage of photons into ordered, intelligent electrical signals that work with our appliances. And even then, you're limited by the availability of sunshine. At night, you're out of luck, so you need energy storage mechanisms. It's easy to store a gallon of gasoline, but it's impossible to store a roomful of photons.

It's in the ordering of energy where the problems lie. When we create ordered energy out of chaos, the unfortunate reality is that we must also create chaos in our wake (I get into this in more detail in Chapter 3). It's not the ordered energy that is the problem; it's the chaos. There are two kinds of chaos left over from the creation of ordered energy; there's heat and there's pollution. The heat is rarely an environmental problem, although when large power plants discharge wastewater into rivers and streams, the heat can alter the natural ecosystem in damaging ways. For the most part, it's the pollution chaos that is causing so much harm to both the environment and human health.

Environmental energy policy directly addresses the chaos side of the equation. Governmental policy has historically been directed toward ensuring a continuous and consistent flow of raw energy supplies into the economy. Secondarily, environmental protection laws have been passed that regulate the way pollution chaos can be released into the environment. There has been very little attempt to bridge the gap between these two disparate ends. Part of the problem has been that there is no collective will among American voters to ramp up environmental protection laws, but that's beginning to change. People are becoming more and more aware that energy chaos is causing not only damaging effects, but permanently damaging effects. It's one thing to live with one's pollution, but it's entirely another matter to insist that your children also live with our pollution.

Getting energy leads to energy consumption

The biggest expenditure of energy on the planet earth, by far, is used to extract and refine energy. When you use an electrical outlet to provide power for a light bulb, you're tapping into a huge interconnected system of coal mines, railroad lines, massive combustion furnace turbines, transformers, high voltage transmission lines, substations, electric meters, and household wiring systems.

The nature of ordered energy is that most of it is wasted (a concept that is explained more fully in Chapter 3). Take light bulbs, for example. Incandescent bulbs convert only 10 percent of their consumed energy into light, and most of this light is wasted. For instance, when you're out of a room being illuminated by light bulbs, the energy is wasted. And you're certainly not taking in all of the light that a bulb puts out, even when you're in a room.

Energy use leads to more energy use. When new, better means of producing ordered energy are developed, new uses are found and the demand goes up. So, the key is for society to find more efficiency in its processes so that people can continue to maintain their lifestyles while at the same time consume less energy.

Reshaping Energy Policy to Include Alternatives

Okay, so there's an infinite source of energy out there, and someone just has to get out there and figure out a way to refine it into useable, ordered energy without too much pollution or negative environmental effects. So what's the problem? Oh, yeah — politicians.

It would probably take an entire book to catalog the money and power associated with the minutia of energy policy. So I'll just leave it at: It's complicated. However, Americans have the ability and know-how to not only reshape the energy policy but also turn the tide for alternative usage. Americans have to come to a consensus and push politicians for alternatives and research. But first, you have to understand the underpinnings of both sides of the debate as well as the availability of alternatives. So the following sections break it all down for you.

Political posturing over energy policy

There is considerable political interest in energy these days, and rightly so. Here's how the sides break down.

On one side are those who oppose a lot of new energy plants and dictate which type of plants can be built and where. Their concern is primarily with the effects that energy consumption imposes on the environment, not the least of which is global warming. And the existence of global warming is a political question.

On the other side are those who believe that Mother Nature can absorb whatever is thrown at her. These people oppose restrictions on new power plants, regardless of the type. They deny the existence of global warming and insist that the earth is in a natural state of temperature flux. In their view, mankind is far too small an effect on the earth to make a material difference.

For my part, I'm very much a political centrist. I find the partisanship and stubbornness of the fanatical fringes of all stripes curious. Here, I take no active position in the global warming debate. Instead, I present data that shows how our current energy consumption is affecting the planet, and I make only a nonpartisan observation that we should be doing better at not affecting our planet.

It's very simple, in my view. Humans leave too much havoc in their wake. Society can have the energy that it craves so much, without having to rely on dubious foreign sources of fossil fuels and the political instability that arises from such. Mankind can use more energy without affecting the environment negatively. I do not push for one solution over another, except that I do push for a wiser course of action. Our energy problems can be solved with alternative energy sources.

What America's policy needs

The historical record indicates that societies grow and flourish with more and more energy consumption and that societies that expand and develop their energy-producing facilities conquer those that don't. This is true militarily, in particular, but it's also true when one considers questions of the basic quality of life (Chapter 1 explains the connection between the evolution of energy sources and civilization).

Our government's policy should be to expand the availability of cheap energy, not to try to shrink it. Restricting energy use will surely result in economic stagnation, and nobody is in favor of that. Yet it seems imperative that the use of fossil fuels be reduced, so what happens then? Reducing fossil fuel consumption does not automatically mean lowering the standard of living. You can transition to alternatives. The path is clear and the technologies, while immature, are ready for market. But there is no record of successful societies reducing their fossil fuel use, so the magnitude of the challenge seems daunting.

Bottom line: Society needs consistent, inexpensive sources of energy in order to maintain and grow. Americans are and have been the most inventive, creative society that ever graced the planet. Because nobody else on the globe is going to lead the way into an alternative-energy future, it's up to America, and America has met incredible challenges before.

The argument for alternatives

The major political thrust today is geared toward ensuring continuous flows of fossil fuels. At the same time, politicians recognize that society's neck is stuck way out when America relies on dubious energy sources. Here is where alternative energy sources come into play in the political picture:

✔ Alternative sources free us from the need to maintain consistent flows of fossil fuels from overseas. This is desirable politically as well as economically. Foreign sources of oil know how badly we need their energy supplies, and they play games and hold us up (successfully) because they know they have the upper hand.

✔ Alternative sources are "distributed," which means that they can be located here and there, on a micro level. If there were enough solar panels distributed around the country, for instance, a major disruption in the flow of fossil fuels would have much less impact because the solar energy sources would still be working. The same is true of nuclear power and biomass. Relying on a small number of huge power sources is bad politics.

✔ A lot of money is sent overseas by way of our trade deficit, and fossil fuels comprise a major component of the trade deficit. Keeping more money at home will help the American economy.

✔ There is a tremendously powerful green movement that lobbies hard in Washington for changes in our energy policies. Alternative energy schemes are very popular with the greens, and rightly so.

Looking at alternative, renewable, and sustainable energy sources

To make sure that resources last, humans need to focus on alternative, renewable, and sustainable energy sources. Energy sources that fail to meet these criteria could eventually be depleted and thus cease to exist.

But what, exactly, do the words *alternative, renewable,* and *sustainable* mean? They're being knocked around quite a bit these days, and you probably hear them quite a bit. All three of these terms are typically associated with energy conservation, even though that is not the case in all instances. The following sections tell you what you need to know about each.

How much does fossil fuel really cost?

Some people argue that a large part of our military is dedicated toward ensuring the consistent supply of fossil fuel energy sources. The solution, in their view, is to somehow include the cost of all of our military and political ventures into the price of fossil fuels. This is a valid point. A lot of money and political capital is spent ensuring the consistent flow of fossil fuel sources, but these costs are not directly assigned to how much is paid for fossil fuels at the pump. Some estimates indicate that gasoline prices would double if the accounting were truly fair. But imagine the political upheaval that would result if that were to happen. Even more to the point, what politician is going to propose doubling the price of gasoline?

Alternative energy

Alternative forms of energy are those that don't include fossil fuels or carbon-combustible products such as gasoline, coal, natural gas, and so on. The origination of the term arises from a need to find alternatives to the hydrocarbon-based combustion processes that now dominate the American economy.

Renewable energy

Renewable forms of energy constantly replenish themselves with little or no human effort. Solar energy is just one example — no matter how much you use, the supply will never end (okay, it may end after billions of years, but your using solar power won't make the sun burn out any faster). Other examples of renewables include firewood, water (via hydroelectric dams), and wind power. The benefits of renewables are that they replenish themselves and so relieve society of its reliance on dwindling, finite supplies. Oil and natural gas, and to a lesser extent, nuclear power, are not renewable because the resources that supply these forms of energy are finite.

Renewables are not necessarily good for the environment. Wood burning, for instance, can be very noxious, particularly when it's done inefficiently.

Sustainable energy

Sustainable forms of energy are not only renewable (see the section, "Renewable energy," earlier), but they also have the ability to keep the planet Earth's ecosystem up and running in perpetuity. The basic notion behind sustainable energy sources is that by their use, society is not compromising future generations' health and well-being. In addition, by using sustainable energy sources, society doesn't compromise future generations' ability to use their own sustainable resources to the same extent that those resources are used now. Who can argue with this very fundamental version of the Golden Rule?

The impetus for renewables

Every year, U.S. coal power plants deposit 100,000 pounds of mercury into the atmosphere, in addition to the two billion tons of carbon dioxide. And each year nuclear power plants produce 2,000 tons of radioactive waste. There are already over 400,000 tons of radioactive waste waiting for some kind of suitable disposal, and the political complexity of finding good solutions has stonewalled all progress.

Most U.S. energy is imported, which channels hundreds of billions of dollars out of the U.S. economy, into foreign nations with dubious and often hostile political systems. The renewable energy sources used in the U.S. are generally locally grown and harvested, which means local jobs and local economic power.

Chapter 3

Putting Together Each Piece of the Energy Puzzle

As you check out the big-picture reality of energy, including its costs, measurements, production, and alternatives, you have to keep in mind some hard facts of energy consumption:

- ✔ **Physics followers:** Regardless of whether the source of energy is fossil-fuel based or alternative, the inevitable laws of physics govern energy production and energy-consumption processes.

- ✔ **Measurements and metrics:** The measurements and metrics are always the same.

- ✔ **Cost of energy:** There's more to the cost of energy than just the cost of the raw fuel itself. You have the costs associated with the machines that extract and refine the energy, the cost of converting the fuel to useable energy, and so on.

When considering energy and how viable alternatives are, you have to take the whole picture into account. This chapter explains how energy is measured (horsepower, joules, and so on), how all energy processes are inherently inefficient, and what the real costs are — tangible concepts that help you understand the issues inherent in all types of energy, including alternatives.

Wrapping Your Mind Around the Metrics of Energy

Oooh . . . scary! Yes, I know "the metrics of energy" may sound frightening, but understanding a few things, such as how energy is measured and types of efficiencies as well as the difference between energy and power can give you a better grasp of the entire energy discussion. I promise — it's *not* scary!

Understanding energy versus power

When you talk about your *energy* bill or your *power* bill, you mean the same thing and use the terms interchangeably. And everyone listening to you as you discuss your energy or power bill understands that you're referring to the same bill because in many contexts, the words *energy* and *power* are interchangeable. Yet the terms mean subtly different things, and understanding these subtle differences gives you a head start to better understanding the basics of energy:

- ✔ **Energy:** The total amount of effort — or work — it takes to accomplish a certain task. To run up a hill, to heat your home, or to dry your hair are all examples of energy.

  Your monthly utility bill is calculated in units of energy. When you pay your utility bill, you pay for energy, not power.

  In the international system (SI), a unit of energy is a *joule*.

- ✔ **Power:** The speed with which energy is being expended to achieve a task. More power means the task is completed more quickly. When your electric meter spins, it's measuring power – the faster the meter spins, the more power you are using. At the end of each month, the total number of revolutions your meter has spun determines your total energy usage. In other words, more power means more energy. Power is calculated by dividing energy by time.

To understand the connection between the energy and power, think about driving a car up a hill. It takes a certain amount of energy to get a car to the top of the hill. How much energy depends on things like the weight of the car and its occupants. A car that is twice as heavy takes twice as much energy to crest a hill. It takes the same amount of energy to drive the car up the hill slowly as it does to drive it up fast (excluding aerodynamic factors we're not interested in here). The more power a car has, the faster it will be able to reach the crest of the hill; a very low-powered car can still make it to the top, it just won't be as much fun.

Common measurements

Engineers describe all energy processes in terms of numbers — it's the vernacular of the industry. For the purposes of this book, we don't need to get into a lot of math and numbers, and I describe most of the energy processes without resorting to high-level descriptions. But you still need a rudimentary understanding of the basic measurement terms and values because numbers are just plain inescapable. Don't worry — just think of this as that special page in the Betty Crocker Cookbook that shows you all of the measurement conversions.

Following are the measurements associated with energy:

✔ **Joule:** The basic unit of energy in the international system is the joule (J).

   $1J = 1 \text{ Kg m}^2/\text{s}^2$ (kilogram meter squared per second squared)

✔ **Btu:** In the English system, the basic unit of energy is the ft-lb, or Btu. A typical home consumes between 50 to 100 million Btus to heat and cool over the course of a year.

   $1J = 0.738 \text{ ft-lb} = 9.478 \text{ Btu}$

✔ **Watt:** Power is energy divided by time, and the standard unit of measurement is the watt.

   1 Watt (W) = 1 joule/second = 3.412 Btu/hr

✔ **Horsepower:** The traditional measurement for power, using horses as the standard. The power with one horse equals one horsepower. The engine in a typical auto has over 160 horsepower; bigger and faster cars have engines upwards of 300 horsepower.

   1 HP = 0.746 kW (kilowatt)

✔ **Calorie:** In physics terms, the amount of energy required to raise the temperature of exactly one gram of pure liquid water by exactly one degree Celsius. One calorie is 4.184 joules.

✔ **1 Quad:** A large amount of energy, equivalent to 10^{15} Btu.

Energy and volume density measurements

The amount of energy that is contained in a unit of fuel (either weight or volume) is called the *energy density*. This is the amount of potential energy available in a given weight or volume of that fuel. Energy density determines how large a storage device is needed or how heavy a fuel will be in that storage device. Table 3-1 lists the energy densities for common raw fuels.

Just how much energy does *that* take?

Here's a list that gives an idea of the relative amounts of energy that are involved in various processes. ***Note:*** The numbers in this table are written in exponential, or scientific, notation, a system that makes representing (and manipulating) very large or very small numbers easier. The number 100 billion (100,000,000,000), for example, appears as 10^{11} in scientific notation; the number 0.001 is written as 10^{-3}.

Sun's total output	10^{31}
Amount that hits earth	10^{22}
World's photosynthesis	10^{19}
Human energy demand	10^{17}
Niagara Falls	10^{14}

Per capita energy use in US	10^{9}
Per capita food energy	10^{7}
Fission of one nuclear atom	10^{-11}
Burning of one carbon atom	10^{-19}

Note how small the last two values are. The reason atomic fission produces so much energy is that the number of atoms "burned" is so huge. In a nuclear power plant there are billions and billions of tiny atomic reactions, which taken together, produce massive amounts of energy. The same may be said for the burning of one carbon atom. In thermal energy processes, massive amounts of atoms are burned.

Table 3-1	Energy Densities in Common Fuels
Fuel	*Energy Density*
Home heating oil	138,690 Btus per gallon
Natural gas	100,000 therms per cubic foot
LPG (liquid propane)	91,690 Btus per gallon
Gasoline	125,071 Btus per gallon
Kerosene	135,000 Btus per gallon
Coal	21,000,000 Btus per ton
Wood	20,000,000 Btus per ton
Electricity	3,413 Btus per kWh (kilowatt-hour)
Hydrogen	52,000 Btus per pound
Enriched uranium	33 billion Btus per pound
Battery	60 Btus per pound

As Table 3-1 shows, hydrocarbons (fossil fuels, like heating oil, natural gas, and so on) are much more energy dense compared to other fuels, and this explains why they're so extensively used. For unit weight, hydrocarbons offer the highest useable energy content, and so the economics favor hydrocarbons. Plus, fossil fuels are generally easier to store and transport than other fuels.

Energy density differences between gasoline and batteries highlight the fundamental problem with electric cars. It takes over 300 pounds of battery to store the equivalent of one gallon of gasoline.

Types of efficiencies

Efficiency is another useful metric. Efficiency is important in every energy process because it describes how much waste is being generated, in relation to the useable work that is being achieved. Some alternative energy processes are not very efficient, but they're still valuable processes. For instance, wood burning is inefficient in terms of how much useable heat energy can be obtained from a given mass of raw fuel. Yet wood burning is a very important part of the alternative energy scenario. Understanding the different types of efficiency helps to understand why some processes are better than others. There are a number of different ways to define efficiency, as the following sections explain.

Energy efficiency

Energy efficiency is simply the ratio of the useful work obtained from a process, by the raw power taken to achieve that process. This is intuitive, but a definition hardens the concept.

An open fireplace, for example, is very inefficient because most of the potential energy that is stored in a piece of firewood goes up the chimney and heats the great outdoors instead of the home space that it's intended to heat. A perfectly efficient fireplace would direct all of its heat onto the skin of the people sitting nearby. An enclosed wood stove, on the other hand, has good insulation and a very hot fire — the efficiency is much greater and the cost of burning wood in a stove is much less, given the same heating capacity.

Automotive fuel efficiency

Fuel efficiency in a car means the amount of miles that can be driven on a gallon of gasoline. A smaller car with a smaller engine gets better fuel efficiency than a big gas pig. The ratio of passenger weight to total vehicle weight is also smaller, which increases the efficiency even more. In fact, the most efficient transport is a bicycle because the ratio of passenger weight to total vehicle weight is nearly one.

Operating efficiency

Efficiency measures can be very specific to a particular system or machine, but they can also reflect broader issues. Operating efficiency is the efficiency of all the individual parts that comprise a whole. The operating efficiency of a home's HVAC system, for example, depends on a number of factors, not just the efficiency of the HVAC system itself: things like the quality of the home's insulation, how leaky the home is, and so on. Due to other factors, the home's overall efficiency may be poor, even if the HVAC is very efficient.

The operating efficiency is the value that has the most meaning because it takes into account all the things that can impact efficiency endeavors. Of course, when operating efficiencies are poor, you can improve the situation by looking at and addressing inefficiencies in the individual components.

Cost efficiency

Cost efficiency is the cost of accomplishing a task divided by the amount of work that is done. This may be the most important efficiency measurement, as it determines how much it will cost to perform an energy process. With some fuels, even though the energy efficiency is high, the cost of the fuel is also high, so the cost efficiency may not be that good. Ultimately, most people are concerned with cost efficiency. The impetus to increase gas mileage for autos has more to do with cost efficiency than energy efficiency.

Cost efficiency and energy efficiency are generally the same thing, but not always. If you're chopping your own firewood from your backyard and using it to heat your home, that's very cost efficient. But if you're burning the wood in a leaky old stove, it's not energy efficient. Burning your wood in a open fireplace is not only energy inefficient, it's cost inefficient as well unless you have a ready availability of firewood that you can cut yourself, in which case the cost efficiency may be very good, although you'll end up doing a lot of work.

Pollution efficiency

Pollution efficiency, a term that's becoming more common, is the amount of work performed by a process divided by the amount of pollution generated by that process. The pollution efficiency of electrical power is terrible. Coal fired furnaces provide most electrical power in the US, and they're dirty. And electrical power from the grid is very energy inefficient as it moves through the grid. But solar PV power, on the other hand, is very pollution efficient.

Pollution efficiency, more than cost efficiency, is the most compelling argument in favor of alternative energy sources. With the economic infrastructure of fossil fuels, most alternative energy schemes are less cost efficient than fossil fuels, but the pollution efficiencies are much better. This will be changing, as taxes and government mandates begin to drive the cost of fossil fuels higher. Carbon taxes are simply a way to converge the cost efficiencies of fossil fuels and alternatives in order to make alternatives more financially attractive. (Carbon taxes are just what they sound like; a tax on any energy source that uses hydrocarbon fuels, and emits carbon pollution.)

The Fundamental Laws Governing Energy Consumption

There are a number of basic fundamentals to all energy processes, and in order to understand why some processes are better than others, it's worth looking into the basics. There is only so much that can be done to improve the energy situation, and there are hard limits. Most people seem to believe that energy efficiency and conservation can lead us out of the wilderness, but in this section I explain why this is going to be a very difficult road to travel.

In this section things get somewhat abstract because the underlying physics is very complex. But the basic ideas can be understood by everybody, so forge on with your chin up.

First law of thermodynamics

Strictly speaking, there is no such thing as using energy. All energy consuming processes follow the first law of thermodynamics, which states that energy can neither be created nor destroyed. It can only change forms. We consume energy, but what we're really doing is using energy in one form (gasoline, for instance) and converting that into other forms.

What we seek when we consume energy is to change the form of the energy into useable work. As that work progresses, the energy changes form once again, usually into heat energy released back into the environment. Massive consumption of fossil fuels releases an equally massive amount of heat into the environment (among other things, like pollutants, but that's another story). Consider these examples:

✔ When we burn gasoline in an auto engine, we emit heat from the exhaust, and we emit even more chemicals than what we put into the gas tank. For each gallon of gasoline (weighing around six pounds) we emit over 20 pounds of hydrocarbons. We heat the air, due to friction when we move. We heat the engine, and that heat is expelled into the environment. The energy of movement converts into heat energy into the environment. While it takes energy to make a car move, it also takes energy to make a car come to a stop. And the energy used to stop the car becomes heat in the brake pads, which then dissipates into the environment.

✔ When we burn wood in a stove, we create heat in the room. While the room may cool down over time, once the fire goes out, the heat does not simply go away; rather, it leaks out of the space where we want the heat. That heat energy that we generate from burning the wood ultimately came from a long growth period for the tree that took sunshine and converted it into wood mass.

✔ Solar electric systems convert energetic photons into electrical energy, which in turn gets converted into some other form of energy depending on which appliances in the house we are powering. The appliances then emit heat into the environment.

✔ Nuclear energy comes from splitting atoms, which generates massive amounts of heat, and that is used to boil water which drives turbines which generate electricity. At each step in the process, inefficiencies generate even more heat.

✔ When we use an air-conditioner, we consume electricity (generated a long distance away) to power a compressor which moves heat from our interior rooms, out into the environment. We use energy to move energy. When we cool a room, we power compressors that generate a lot of heat in the process. All that heat escapes into the environment, along with the heat we remove from the room.

All fossil fuel is burned, or combusted. Millions of years ago, life forms on the planet died and collected in deposits that either solidified (as with coal) or liquefied (as with crude oil) or gasified (as with natural gas). The vast majority of our energy needs are met with fossil fuels, and these are all burned so as to generate heat. We convert this heat into useable work, or ordered energy, and then when we use that ordered energy, the vast majority of the energy is converted back into heat.

Second law of thermodynamics

The second law of thermodynamics states: disorder of any closed system can only increase — in plain English — waste is unavoidable. The physics term for disorder is "entropy," and the mathematics and physics are very complex, so I will not get into it here. Think of disorder as nothing more than chaos.

The second law of thermodynamics is every bit as important in the scheme of energy consumption. It states that the disorder (or *entropy*) of any closed system can only increase, or, to put it more simply, every energy conversion process produces at least as much waste as it does useable energy. This implies a limit to how efficient we can make energy consumption processes.

Examples of the second law in action

At each stage of an energy generating process, more energy is wasted than gets passed on to the next phase of the process. As we invent more sophisticated machines, we need more and more refined energy, and so the problem of waste only grows with our increasingly complex information economy. A very sophisticated machine wastes more energy than a coarse machine. A high speed computer wastes hordes of energy; an abacus very little. Bottom line? There is no such thing as 100 percent efficient because all energy consuming activities are wasteful. This is not a cynical comment on humans' inability to create efficiency, it's a physical fact. In a perfect world, there will be a lot of waste.

Here are some examples of the second law at work:

- When we generate electricity, we will always end up throwing away at least as much energy as the electricity we are creating.

- In the process of driving a car up a hill, a certain amount of energy is expended. At least that much energy is also wasted. In fact, more is wasted due to inefficiencies.

- The huge power plants that utilities employ to generate electrical energy discard a sizable proportion of the raw energy consumed by way of wasted heat into the environment. Smaller power plants are even worse. See the section "Why bigger is better" for details.

Waste may not be desirable, and we would like to minimize it, but there's only so much we can do. We cannot violate the laws of physics, and the second law of thermodynamics implies a limit to how well we can do. This is not to say that we have to give up the game. What we must do is get smarter about what we are doing with all the energy we consume and about what types of raw energy we are consuming. In particular, if waste is unavoidable, we need to mange that waste much better than we have been doing.

Looking at efficiency pyramids

Humans seek ordered energy, not simply energy (refer to Chapter 2). In the process of ordering energy, a huge amount of energy is tossed aside. This is as inevitable as the sun coming up tomorrow. As per the second law, it is by tossing aside energy that we achieve the ordered energy that we seek.

Compare a laser system to sunlight. Sunshine is a mad rush of photons of all wavelengths flying willy-nilly through the air; a laser's light is the exact opposite. In lasers, which are highly focused beams of light, the photons are in lock-step with each other. They all match in frequency, and they're all in phase, which means the individual wavelengths are aligned. The energy of a laser beam is as dense as the core of a nuclear reactor. With sunlight you can grow grass in a pasture (a nice enough thing to do, but it does have limitations). With lasers you can shape a human eye with striking precision.

But a laser is also a tremendous waste because it takes so much energy to create those highly coherent photons. The same limitations apply to computers, automobiles, and several other products.

Transport, in general, involves a huge waste of energy. Figure 3-1 illustrates the pyramid nature of the wasted energy process in building an automobile. Note how much energy is consumed in the factory that builds the auto. This represents the *invested energy*, the energy used to manufacture a product (in the literature, you may see this referred to as *embodied energy*, or grey energy, but it all means essentially the same thing). When the auto is moving, the vast majority of energy expended goes toward moving the machinery, as opposed to the occupants. So moving a passenger — represented by the thin spire at the top of the pyramid — required all the energy expended for refinement, manufacturing, and moving the car itself. If you consider that the passenger could have moved much more efficiently under the power of his or her own two feet, you can see that transport, in general, entails a huge waste of energy.

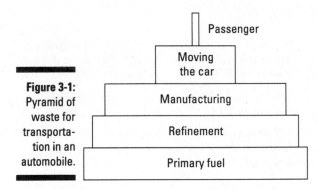

Figure 3-1: Pyramid of waste for transportation in an automobile.

Pyramids may be drawn for any energy consuming process, and the graphical nature brings to light the second law of thermodynamics: that the more refined or controlled the energy process, the more waste is involved.

Carnot's law

At the turn of the century, steam engines were the predominant form of energy generation in our economy. The fact that steam engines required a lot of wood and water necessitated an efficiency analysis to find ways to lower costs while producing the same power levels. As a consequence, a French engineer named Henri Carnot came up with a very important law which governs all combustion machines. In a nutshell, Carnot's law states that the maximum efficiency of any thermal energy process is determined by the difference in temperature between the combustion chamber and the exhaust environment where the waste is channeled.

Now, I didn't just choose this particular law out of a physics hat; instead, as the following sections reveal, Carnot's law helps to explain the efficiency of larger versus smaller power plants as well as finally explain why you pay peak and off-peak rates for power.

Practical implications

According to Carnot's law, it's the temperature difference that matters, not the temperature of the actual combustion process. According to Carnot's law, the higher the temperature of the combustion process, the better. But that's only half the story. The exhaust temperature is just as important. When the outdoor temperature is high, the efficiency of a combustion process also goes down because the difference in temperature between the combustion chamber and the ambient is less.

Here are the practical implications of Carnot's law:

- A steam engine working in a cold climate is more efficient than in a hot climate because the exhaust ambient is cooler. This is true even if the combustion temperature is the same in both climates.

- A steam engine with a hotter burning fire is more efficient; for that reason, well insulated combustion chambers work more efficiently.

- The most efficient engine of all is the jet turbine because the fuels are burned at tremendously high temperatures and pressures and the exhaust temperature is very cold.

- Residential natural gas heaters work better when the gas is burned at a hotter temperature.

- Wood stoves work much better than open pit fireplaces because the burn temperature inside of a stove can be much greater than in an open pit.

Why bigger is better

Within the context of alternative energy solutions, the Carnot cycle is important because it explains why larger power plants are more efficient than smaller ones.

Coal fired power plants burn raw coal to boil water, which is used to turn huge turbines that create electricity. As per Carnot's law, the hotter the coal is burned in relationship to the exhaust temperature the more efficient the plant operation will be.

Consider the size of a burn chamber. Each chamber has a volume and a surface area. The surface area causes inefficiency because heat escapes through the surfaces, into the surrounding ambient. When heat escapes through the surface, as opposed to being used to generate steam, the internal temperature goes down and so there are two effects which lower efficiency: escaping heat and lower internal temperatures.

Imagine the burn chamber as a large sphere. The larger the diameter of the sphere, the higher the internal temperature can be because the ratio of volume to surface area is proportional to the diameter of the sphere. With a larger sphere, the inefficiencies are much smaller. Less of a percentage of heat escapes through the surface area, and the internal temperature can be higher because the temperature will be the hottest at the central core.

This is the key reason why power plants are built in massive scales (other, less important reasons include things like transport efficiencies of the raw materials and wastes). These power plants can be located far from urban areas, so they are out of sight.

The problem with "peak power" times

Summer peak times are a utility company's nightmare. Peak power consumption occurs on hot summer afternoons. This is when the utility is called upon to provide the maximum amount of electrical power because air conditioning uses electricity. At peak times, businesses and offices are all using their air-conditioners, and homes are also cranking away. (Peak power issues do not occur in the winter, when it's really cold, because a large percentage of heating energy comes from non-electrical sources, like wood, natural gas, etc.)

When the utilities cannot meet power demand with their massive power plants, they need to call upon smaller utility providers, and many of these are little more than huge diesel engines connected to generators. According to Carnot's law, these smaller power generating machines can never achieve the efficiencies of the massive power plants, and this means the cost of the power is necessarily higher. And operating at peak when the outdoor temperature is high means that the combustion process is even less efficient.

So the utilities take a big economic hit when they need to call up the backup power plants. This is the impetus for peak and off-peak rates, more commonly called time-of-use, or TOU. A special meter is installed at a user's location which can measure power consumption at peak times (usually noon to six PM) and off peak times. The utility charges different rates for these times, often as high as a three to one ratio. TOU metering serves to decrease the consumption at peak times, which is good economics for society. In the future, it is likely that all new buildings will be outfitted with TOU meters.

Understanding Electrical Energy

The vast majority of electrical energy is produced by burning fossil fuels, a topic I cover in detail in Chapter 5. And most of the energy used in the commercial and residential sectors is electrical energy. For that reason, it's of value to look at how electricity works, and some of the numbers that describe the process. Regardless of whether conventional energy sources are used, or alternative, the physics of electricity is equally important.

A short primer on electricity

All atoms have electrons, which have negative charges (unlike the protons, another part of the atom, that have positive charges). When the electrons can break free, or detach, from their atoms and move through a substance, whether solid, liquid, or gas, you get electricity.

How freely electrons move depends on how easily they can break away from their constituent atoms. In materials like cotton, wood, plastic, and glass, the electrons are stuck; they can't move around because they are captured by the atom that they belong to. These materials are *insulators*. In metals, on the other hand, the electrons are only slightly held to their host atoms and are able to move around. Because these materials enable the flow of electricity, they're called *conductors*. (Some materials, like semiconductors, which are at the heart of all electronics devices, have properties that are somewhere between metals and insulators. The electrons can flow, but only under special conditions.)

The electrons flow through a conductor if something spurs their movement. That something is a voltage. When a voltage, or electromotive force, is applied to a material, the electrons flow depending on the amount of *resistance* the material has. Resistance is just a measure of how easily the electrical charge flows through a substance. Metals have low resistances, whereas insulators feature very high resistances. As you can imagine, resistance causes inefficiency. When the resistance of a material is very high, it takes a lot of energy to move the electrons through that material.

Producing electrical power efficiently and safely

The challenge power plants face in making the power they produce available to customers is how to maximize efficiency and at the same time control the voltage so that it can be used with some degree of safety.

Formulas for the math minded

There are a number of formulas that let you calculate electrical power. In reading these formulas, keep in mind these definitions:

Voltage (V): The electrical "pressure" between two points on a circuit. When the voltage is very high, there is a strong force that encourages the flow of electricity, or the flow of electrons to be more precise.

Current (I): The amount of electrical flow, or the sum total of all the electrons which are flowing.

Resistance (R): The resistance to electrical flow.

Here are the mathematical relationships between voltage, current, and resistance:

Formula	Translation
$P = VI$	Electrical power is equal to Voltage times Current. This formula shows that you can raise the voltage and reduce the current and end up with the same result
$P = V^2/R$	Power equals voltage squared divided by resistance
$V = IR$	Voltage equals current times resistance
$P = I^2R$	Power equals current squared times resistance

Power transmission wires are resistors — as the previous section explains, all electrical paths have resistance of some kind or another, there's no escaping it. As the power travels through the transmission wires, the power dissipates. One way to reduce the power dissipation is to get the current that travels through the lines as low as possible.

Because electrical power is equal to voltage times current ($P = VI$), it's possible to raise the voltage and reduce the current and still end up with the same amount of power. In reducing the current, the line losses are reduced as is the amount of power wasted in the transmission process. (*Note:* All these ideas can be expressed mathematically, but you don't need to know the formulas to understand the general concepts; if you're interested, though, pop to the sidebar "Formulas for the math minded.")

Therefore, it is very desirable to use extremely high voltages on transmission lines. The result is transmission lines that commonly have voltages over a hundred thousand volts. That's all well and good, but household voltages are 110VAC. To solve this problem, you need transformers to convert the high voltages down to the relatively safe and manageable household voltages.

The electrical grid is a huge complex of power transmission wires and transformers that alter the voltage/current characteristics of the power being delivered to utility customers. All of this is done to enhance the overall efficiency of the entire system. In essence, with all electrical systems, the higher the voltage that can be used, the lower the inefficiencies. Solar PV panels, for instance, are configured to create the highest voltages possible if there is some length of wire between the solar panels and the end user. This effect is also why large appliances use higher voltages – because they use so much power, it is highly desirable to reduce the wasted energy in line losses as much as possible.

Tracking the Real Cost of Power

Until the steam engine came on the scene, most energy was expended on growing crops and feeding animals. And up until 100 years ago, most energy was harvested at the surface of the earth, at great hardship and labor. The costs of obtaining raw fuel were much greater than the costs of the equipment used to burn that fuel. There were no sophisticated machines that performed magical feats by consuming ordered energy. People's energy needs were much simpler: obtain heat and cook food. But that's changed drastically.

The real cost of power is impacted by a number of things: the cost of raw fuel, the cost of converting that fuel to useable energy, the cost of the equipment and its upkeep, and so on. The following sections explain how these factors affect the real cost of power. What you'll discover — perhaps to your surprise — is that raw fuel costs actually play a relatively small part in the overall cost.

Raw fuel costs

In recent years, the American economy has spent around $400 billion per year on raw fuel. With oil costing $100 a barrel, this number has increased, but it still represents less than the total capital outlay dedicated to producing and consuming energy.

Taking a look at the relative costs of different types of raw energy is illuminating. (By raw, I mean the cost at delivery; this doesn't take into account how the energy is actually used, or how much the equipment costs to convert the raw energy into useable form; see the next section for that). Relatively speaking, electricity is the most costly, nuclear energy is the cheapest, and a bunch of other types of energy fall somewhere in between (see Table 3-2).

Table 3-2	Comparing Raw Energy Costs
Type of Energy	*Cost (in $) per Million Btus*
Electricity	29.3
Liquid propane gas	18.54
Gasoline	15.19
Kerosene	11.11
Heating oil	10.82
Natural gas	10.00
Coal	9.52
Wood	7.50
Uranium (nuclear)	0.00033

As Table 3-2 shows, electricity is by far the costliest energy source. This is because the *power transmission grid,* the huge matrix of power wires and transformers used to get the power from the utility generators to your home, is so big and unwieldy. Think of it this way: When you use electricity from the grid, the production chain begins in a coal mine, then the coal must be transported to a power plant, and then power is generated in huge, expensive machines. This electrical power gets transmitted over long lengths of wire until it finally reaches your home. Several companies are involved — a coal-mining company, a transport company, and a utility company with linesmen, managers, secretaries, lobbyists, and lawyers — not to mention that a whole host of governmental regulations have to be met, and taxes and tariffs need to be paid.

Even though electricity is the most expensive energy source, it's used almost exclusively in a majority of homes. Why? Because it's the most convenient way to use power. It's the most ordered; you can get it to do most anything. All you have to do is plug your toaster into a wall socket and voilà! You don't need a storage tank (like you do with propane or fuel oil), and you don't have to worry about the danger of explosions or flames (although you can get a shock without even trying too hard). You don't need to go into the woods with a chainsaw and cut down a tree, and you don't need to stack electricity on the side of your house and keep the spiders out of it. Best of all, the utility company is responsible for maintaining a continual flow into your home.

Looking at conversion costs

Most people are familiar with their gasoline bills and their utility bills, but raw energy is always consumed in a machine of some kind (unless you're burning firewood in a hole in the ground for your heat source, and even then you need matches, or a flint stone).

Raw energy costs, explained in the preceding section, are only part of the picture. Most energy sources need to be burned or combusted on-site in order to extract their potential energy. Some conversion methods are more efficient than others. The list in Table 3-3 shows the same fuels as those listed in Table 3-1, but this time with all the various production factors that come into play.

Table 3-3	Comparing Actual Energy Costs
Type of Energy	**_Cost (in $) per Million Btus_**
Gasoline	75.96 per million Btus
Electricity	29.30
Liquid propane	23.18
Coal	15.87
Kerosene	13.89
Heating oil	13.52
Wood	12.50
Natural gas	12.05
Uranium (nuclear)	0.024

Gasoline is now the costliest energy source because it requires the most complex types of machinery to use. Wood is a cinch to burn; you don't even really need a machine, but if you do use one, like a wood stove, the machinery is simple and effective. If you cut your own wood, the cost is even lower (notwithstanding all the tools and food you need to accomplish the task).

Capital equipment costs

The $400 billion a year we spend on raw fuel doesn't tell the whole story. We also spend over $500 billion per year on new capital equipment dedicated to refining, storing, and transporting fuels. Add the value of existing capital equipment, and the maintenance and financing costs associated with it, and total expenditures on capital equipment exceed $700 billion per year. And much of this capital equipment cost is dedicated to equipment that is used to extract and refine the raw fuels.

Grid power costs around 12 cents per kWh (kilowatt-hour), although this fluctuates quite a bit from region to region. The same amount of energy contained in a chunk of raw coal costs only a third of a cent. Computer grade power, that is, power which is highly refined and consistent, costs over $3 per kWh. Yet the consumption of this high grade electricity is the fastest growing sector of the US economy, and the majority of value added to each successive increment of progressively more refined energy comes from the equipment used for the refinement. In short, that original lump of coal for a third of a cent gets turned into high-grade power for $3 per kWh.

Energy utilization

For every source of energy, the cost of the production plant is directly proportional to the plant's capacity (the larger the plant, the higher the capital costs), while the revenue is proportional to the energy generated. Plant utilization factor is the ratio of energy produced to the total capacity of the plant if it were in full up operation 24/7. Table 3-4 lists some utilization factors for common fuel sources.

Table 3-4	Utilization Factors for Common Fuel Sources		
Source	*Used (EJ)*	*Capacity (GW)*	*Utilization factor*
Thermal	9.69	600	51.1 percent
Nuclear	2.71	98	88 percent
Hydro	0.97	99	31.1 percent
Wind	0.021	4.28	15.4 percent

The small utilization factor for wind is due to the variability of wind. Solar suffers the same problem due to sunlight availability.

The net economic effect is to make investments in low utilization technologies riskier, with a more difficult payback. Nuclear utilization is so high because nuclear plants are used constantly. They cannot be turned on and off at will, and the capital equipment costs are so high that when a nuclear plant is commissioned, it's used as much as possible.

Putting it together in practical terms

For an automobile, fuel costs represent less than 20 percent of the total cost of driving. The 5 percent gas guzzler tax that the government imposes hardly dissuades drivers from driving gas guzzlers because the raw fuel/capital equipment percentage is even less than 20 percent for expensive cars.

The difference between cost and fuel is even more stark in the remainder of the energy economy, which is much less dependent on fossil fuels. The non-transport sector already gets 90 percent of its energy from electricity, and the price of electricity depends even less on the cost of raw fuel than transport.

Oil generates less than 5 percent of the domestic electricity production. The costs of coal and uranium don't factor into the price of electricity either, because the greatest component of the electrical cost is in the massive power plants and distribution systems, as well as the labor costs in the huge utilities that administer the power to our homes. Raw fuel costs represent half the delivered cost of electricity from gas fired turbines, around one third of coal-fired power, and only 10 percent of nuclear power. Solar and hydro-power have no raw fuel costs at all. Yet all these sources are not too far apart in terms of delivered cost.

In the past 50 years, hardware has grown more efficient, and so even as raw fuel costs fluctuate and increase, the real cost of electricity has steadily fallen. When electricity prices gyrate, it's because of governmental interference in markets, not because of fuel costs or hardware issues. In the long term, the real cost of energy has dropped steadily, and this is why Americans use more and more energy every year.

Part II

Digging Deeper into the Current State of Affairs

The 5th Wave   By Rich Tennant

One of the things we do to conserve energy in the house is to make sure all the lights are turned off before we go to bed.

In this part . . .

Fossil fuel consumption has both advantages and limitations, and in this part I help you understand what those are. In Chapter 4, I describe how the world, particularly the United States, consumes fossil fuels. In Chapter 5, I explain the myriad ways that fossil fuels are used and the machines that convert fossil fuel–potential energy into useful work. Chapter 6 addresses conservation and efficiency, which, to some people, are considered the most important alternative-energy sources. I wrap up the part in Chapter 7, by getting into the details of why fossil fuels are causing so much havoc in our society.

Chapter 4

Developing a Snapshot of Fossil Fuel Use and Availability

. . . .

In This Chapter

▶ Taking a glance at energy sources

▶ Checking out energy consumption in the U.S. and around the world

▶ Getting a glimpse of the cost and availability of fossil fuels

. . . .

*W*e are addicts.

Our drug of choice? Fossil fuels. And it's not just the U.S., but the entire world that shares this addiction. The vast majority of the energy infrastructure established over the last 100 years relies on fossil fuels, and so for now the world is driven by thermal combustion processes generated by fossil fuels, which are relatively cheap and abundant. While the situation isn't likely to change any time soon, it will change eventually (just see the other chapters in this part).

This chapter takes you on a quick stroll through the energy sources currently available, explains consumption trends (you'll see that fossil fuels are currently the fuel of choice for just about every economy), and examines the current availability and cost of fossil fuels.

A Quick Look at Fuels from the Fossil Fuel Era and Beyond

Under huge pressures and amidst the Earth's core heat, decayed and decaying organic materials formed complex compounds primarily composed of hydrogen and carbon atoms. These complex, carbon-based compounds are more commonly referred to as *fossil fuels*. The three most commonly available fossil fuels are coal, petroleum, and natural gas.

The following sections outline basic information about each type of energy — both fossil and nonfossil fuels — as well as an overview of the world's use of these energies. You can find more detailed info about fossil fuel sources in Chapter 5 and the chapters in Parts III and IV are devoted to explaining alternative energy sources.

The evolution of energy use

Humans stumbled across the potential in fossil fuels when they discovered fire in prehistoric times, but the modern fossil fuel era didn't get into full swing until the Industrial Revolution in the 1700s. It was then that coal supplanted wood because of its superior energy density (it packs more energy into a given volume and weight than wood does). Coal consumption peaked as the demand for oil and natural gas increased. Figure 4-1 shows how the world's use of different fossil fuels evolved since 1800.

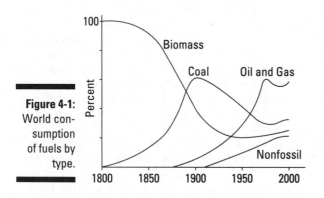

Figure 4-1: World consumption of fuels by type.

Hydropower, represented by the "Nonfossil" line in Figure 4-1, came on the scene around 1920, and that explains the increase of nonfossil energy sources since then. The further increase in nonfossil fuels after around 1970 is due to nuclear power plants and other alternatives.

Petroleum products

Most petroleum products, which include gasoline, propane, and heating oil, are derived from crude oil drilled from beneath the ground. The energy that derives from petroleum products comes from burning the fuels, or *oxidizing* the fuels to be more precise. Carbon dioxide, the molecule responsible for global warming, results from the burning of petroleum sources. Petroleum products account for 39 percent of all the energy used in North America.

In order to be converted into gasoline and other useful fuels, crude petroleum must be "cracked," which is a process akin to a distillery where different boiling points of a complex liquid are reached and the constituent materials boil off separately. Petroleum refineries are huge chemical processing plants where crude oil is cracked.

Following the cracking process, various fuels and other petroleum products are further separated. Gasoline is the most prevalent fuel produced from petroleum. The amount of gasoline derived from a barrel of oil is dependent on the quality of the oil, but roughly 25 gallons can be refined from a barrel (which is 42 gallons) of crude petroleum.

Petroleum refining is the most energy intensive manufacturing industry, accounting for over 7 percent of all U.S. energy consumption. It takes a lot of energy just to supply energy.

Natural gas

Natural gas is drilled from beneath the surface of the earth, just like petroleum, but it comes in gaseous form instead of liquid or solid. Transport is problematic; so is storage. Most natural gas systems rely on a vast matrix of interconnected pipes and pressure regulators. Natural gas is more common in urban areas than in rural areas because of the need for this matrix of supply lines.

Natural gas is generally favored by environmentalists because it burns cleaner than petroleum products and coal. Because availability of natural gas is good in the U.S. (proven domestic reserves are very large), natural gas consumption has increased as well. Most new power plants are being fired by natural gas, and many homes and businesses are converting their old coal and oil burning heating systems to natural gas. The U.S. extracts and markets around 20 trillion cubic feet of natural gas annually. Natural gas makes up 24 percent of total energy consumption, with very little of that going toward transport because of safety reasons.

Coal

Coal is mined from the ground and comes in a solid, blackish, oily form that's easy to burn but very difficult to burn cleanly. Coal is the most common form of energy in North America because there's so much of it available domestically and it's relatively easy to extract from the ground, compared to petroleum.

While crude petroleum consumption has increased dramatically over the last 20 years, coal consumption has increased even faster. Between 1980 and 2005, coal production in the U.S. increased by around 40 percent and the rate of increase remains consistent.

Most of the U.S. electrical power is derived from coal combustion, and this causes a lot of the impetus to go to alternative energy because coal is so dirty. There are methods being developed to make coal burn cleaner and release less pollution into the environment, but even when these advances are perfected, coal will still be a major culprit of greenhouse gases. There is simply no way to burn coal cleanly.

Because the U.S. has such a large domestic supply of coal, it is an attractive solution politically because supplies won't be disrupted by instability in other countries, conflicts with other nations, or global environmental or political crises. Coal also provides a lot of domestic jobs.

Nuclear

The same atom-splitting physics that provides us with those wonderful nuclear weapons is used to heat water, which is then used to turn large electrical generating turbines. Much of the controversy about nuclear energy relates to nuclear waste products, which are very difficult to deal with. Plus the word "nuclear" is associated with atomic weapons, which nobody likes.

Nevertheless, nuclear power is coming back into favor because it doesn't generate the carbon dioxide pollution blamed for global warming and other problems. Plus the known reserves of uranium (used in the nuclear reaction) are huge compared to fossil fuel sources.

In some countries, like France, nuclear power provides a majority of the electrical power. Yet it comprises only 11 percent of North American energy consumption. Look for this percentage to increase dramatically in the next couple decades. However, this will take some time because nuclear power plants are so expensive and difficult to build with all the safety features required. Nuclear plants also require years of approvals and planning, although there is political pressure now to relieve this burden.

Hydropower (dams)

Tremendous water pressure builds up at the bottom of large dams located on river ways, and this pressure is harnessed to turn turbines that generate electricity. Hydropower is very clean, but the problem is that dams affect the river's ecosystem. Salmon runs are obliterated and the natural beauty of the river is permanently altered, almost always for the worst (unless you like to water-ski).

Hydropower accounts for 3 percent of North American energy consumption, particularly in the American Pacific Northwest. It accounts for around 8 percent of electrical energy production, and that's significant because most of our electrical energy comes from burning coal, which is very bad for the environment (see the section, "Coal," earlier in this chapter).

Biomass (firewood and other natural sources)

Leaves falling from the trees in autumn are biomass, so are cow manure, horse manure, hay, weeds, corn husks, and so on. Firewood is the most common form of biomass used for energy. Total biomass energy production accounts for only 3 percent of North American energy usage (most of it for heating homes). Because of its wide availability, biomass is seen as an excellent sustainable energy source.

When biomass is left to rot on the ground, it releases just as much carbon dioxide into the atmosphere as when it's burned. So biomass is a very attractive alternative-energy source because burning it doesn't release any more carbon dioxide than is released naturally through the process of decay.

Geothermal

The air you breathe and the ground beneath your feet provide the source for geothermal energy. Although the physics is complicated, you can accomplish heating and cooling by using heat exchange mechanisms that move temperature differences from one place to another. If you want heat in your home, you take some from the air or the ground and channel it into your home. If you want to move heat out of your home, you channel it back into the environment.

Geothermal energy provides 0.3 percent of North American energy needs.

Wind

Everybody has seen windmills. Traditionally they have been used to pump water from the ground (you see them in old western movies at nearly every ranch). But modern windmills produce electrical power as well. Wind power only provides 0.11 percent of energy needs, but this is changing fast because wind is a very clean energy source, and improvements in the technologies have been impressive. The problems with windmills include:

✔ They're unsightly.

✔ They make noise.

✔ They may kill wildlife, particularly birds.

Wind power is a form of solar power because wind is caused by thermal temperature differences between regions of the globe.

Solar

The sun radiates a tremendous amount of energy — 35,000 times as much energy as humans use — onto the surface of the earth, pollution free. Unfortunately, tapping into this energy isn't simple. Nevertheless, people have figured out ways to use solar energy to heat water and to produce electricity. There is a tremendous push to increase solar production, and many governments are subsidizing solar equipment in both residential and commercial settings. Right now solar power accounts for only 0.06 percent of energy production (if you don't include all the natural light and heat the sun produces), but that's changing fast.

Electricity

Of all the energy sources in use, electrical energy has grown the most and has also grown into the most important energy source. Electrical energy is generated through processes powered by other energy sources. For instance, coal is burned and water is boiled and the pressure is used to turn huge turbines that generate electrical power. In North America, electrical energy comes primarily from these sources: coal, nuclear, and gas. In the following sections, I list some of the disadvantages and environmental impacts that generating electricity causes whether it's generated through alternative sources or not.

One of the key reasons that electrical energy consumption has grown so much is the increased use of computers. As the complexity of computers increases, so do the power requirements, and in particular, highly ordered power. Fast computers with hordes of memory require a lot of power supply filtering, which entails waste and inefficiency.

The dirty truth about electricity

Although electricity is typically considered clean and environmentally friendly, it's often generated through processes that rely on "dirty" energy sources. Even those processes that use renewable energy sources have their own problems. When considering how "clean" electrical energy is, you need to take into account the source of the electric power, as well as the inefficiencies inherent in those particular systems.

While clean once it reaches your home, electrical energy has most likely been generated by a combustion process, so it is not necessarily environmentally benign. Consider, for example, that about 90 percent of the coal produced in this country is used to produce over 50 percent of our electricity. Due to its high carbon and sulfur content, burning coal is extremely dirty, and coal byproducts are the leading contributor to global warming (nearly 1 billion tons of carbon dioxide are produced each year in the United States alone; China has surpassed this). Mercury is another byproduct of burning coal; it pollutes air, land, and water. Electrical power generated by coal-burning processes, therefore, is a very polluting source of energy.

Disadvantages with electricity generated by alternative means

Don't think, though, that electrical energy generated by alternativeenergy sources is problem free because it's not:

- ✔ Wind generators cover open hillsides with noisy, ugly windmills. Birds and bats get killed, sometimes in large numbers. On the other hand, windmills are pristine compared to the alternatives, so criticisms should be taken within context.

- ✔ Hydroelectric generators dam up rivers and affect such diverse phenomena as salmon runs and the plants and animals that live along the riverside. Dams also create a lot of underwater decay in the upper reservoirs that releases carbon dioxide into the atmosphere.

- ✔ Geothermal wells release arsenic into the environment.

- ✔ Nuclear power plants produce horrendous wastes, and people have yet to figure out a good way to get rid of these byproducts.

- ✔ Power lines emit radiation, heat the air, buzz, and catch birds and planes. Their potential for danger is even greater when they're downed.

Energy Consumption Trends

By checking out how countries around the globe are using energy and in what form, you get a clear picture of what fuel sources the globe is relying upon. To paint such a picture I provide you with, in the following sections, an overview of how the globe uses energy and in what forms, comparing countries' usage as well as noting how increased industrialization drives energy consumption.

Global energy consumption and use

In 2002, the world used 409 Quads of energy. Where did all that energy come from? Table 4-1 lists the power sources and percentage of the total usage each represents, both globally and in the United States.

Table 4-1	Power Sources, in Percentages, in 2002	
Energy Source	Percentage of Total Usage Globally	Percentage of Total Usage in U.S.
Petroleum	39 percent	21 percent
Natural gas	23 percent	27 percent
Coal	24 percent	32 percent
Hydroelectric	6 percent	3 percent
Nuclear	7 percent	11 percent
Renewables	1 percent	3.5 percent

As Table 4-1 shows, the combustion of hydrocarbons (fossil fuels, like petroleum, natural gas, and coal) comprised 86 percent of the world's energy consumption — a number that has grown in the last decades; as China and India come on line, it will continue to grow — and 80 percent of the U.S. consumption. Hydroelectric, nuclear, and renewable fuels account for the alternative contributions, and amount to only around 14 percent of the world's energy sources — the U.S. does slightly better with alternatives, coming in at about 18 percent. *Note:* This percentage does not include alternative sources like solar room heating (that is, using direct sunlight for warmth) and bicycle riding, however.

And where did all this energy go? Table 4-2 gives a sampling of countries. From this table you can see that the U.S., despite the fact that it represents only 4 percent of the world's population, consumes about 24 percent of world energy consumption and uses about 25 percent of the world's oil.

Table 4-2	Comparison of Energy Use around the World			
Country	Percentage of World's Population	Percentage of World's Energy Consumption	Percentage of World's Oil Consumption	Consumption per Person per year (MMBTU)
USA	4.6	23.7	25.3	339
Canada	0.5	3.2	2.7	418
Mexico	1.6	1.6	2.5	65
Western Europe	7.8	17.6	19	149
India	17	2.8	2.8	13
China	21	10.5	6.6	33
Japan	2	5.3	6.8	172

Some interesting things to note about the data in Table 4-2: Canadians use more energy per capita than do Americans. True, Canada is cold, and its citizens require a lot of heat. But their high consumption also reflects the fact that energy is relatively inexpensive in Canada. When energy is cheap, there's little incentive to conserve or practice efficiency and conservation.

What's more interesting is to compare per capita consumption in the U.S. (339) with per capita consumption in Western Europe (149). Western Europeans, in many ways, enjoy better lifestyles than Americans. So why is their per capita consumption so low? Because they've been inculcated by high energy prices for so long that energy conservation and efficiency are ingrained into the very fabric of their societies. Western Europe provides Americans with a paradigm; this is where we are heading, and it's not bad at all. In fact, there may be some interesting but subtle increases in our quality of life when we end up using less energy.

Energy consumption in growing economies

There is direct correlation between the per capita usage of energy in a nation and its corresponding GNP (Gross National Product). As less developed countries, interested in catching up economically with the world leaders, industrialize, their energy consumption increases. As the Chinese and Indian economies, for example, require more energy to power their development, global energy usage will increase. Today this growing demand is being met primarily through fossil fuels.

Efficiency may alter this trend toward more energy usage, but until efficient solutions are economically feasible, many developing countries will continue to rely on relatively cheap fossil fuels for power, despite the environmental costs.

The United States is achieving higher levels of GNP with less energy, but not because of efficiency and conservation. Most of our heavy industries (like steel and manufacturing), which use a lot of coal-fired power, are moving overseas.

The U.S. is more efficient at converting its energy use into Gross Domestic Product (GDP) — this is a metric known as *energy intensity* and is important in discussing economic issues related to energy. In 1950, the U.S. consumed 20,000 Btus for each dollar of GDP, but in 2000, that figure was roughly half, which implies that the U.S. is getting much more efficient in at least one way: The U.S. may be an energy pig, but it also provides a lot of value for the energy it consumes, providing the world valuable economic output at a relatively small energy cost.

Energy consumption in the U.S.

The United States is the world's largest consumer of energy, and the disparity in energy use per capita is growing. In 1950 the U.S. was energy independent, consuming only 35 Quads of energy. Fifty years later, the U.S. population had increased 189 percent while energy consumption grew 280 percent. While U.S. energy consumption overall has grown, the energy sources the U.S. relies on have changed, as Table 4-3 shows.

Table 4-3		Fuel Sources in the U.S. since 1925		
Fuel	*1925*	*1950*	*1975*	*2000*
Wood	7 %	3	0	0
Coal	65	37	24	32
Oil	19	39	38	21
Natural gas	6	18	32	27
Total fossil	97	97	91	81
Nuclear	0	0	7	11
Renewable	3	3	3	8
Percentage into electricity	0.5	1.5	15	38

As Table 4-3 shows, electrical production, regardless of the fuel type, increased sharply between 1925 and 2000.

Table 4-4 shows the four major sectors of the U.S. economy that use energy and the amount of oil consumption that each sector uses. Note that transport uses a majority of oil while representing only around a quarter of the total energy consumption.

Table 4-4	U.S. Oil Consumption by Sector	
Sector	*Percentage of Total Consumption*	*Percentage of Total Oil Consumption*
Commercial	18	2
Residential	21	5
Transport	27	65
Industry	34	25

Table 4-5 shows projected fossil fuel energy consumption for the United States, in relation to the world. It used to be that Americans consumed far more energy than the rest of the world, but this projection shows how the rest of the world is catching up to us and adopting our ways.

Table 4-5	Projected Fossil Fuel Consumption			
Source	*2000 (U.S.)*	*2025 (U.S.)*	*2000 (World)*	*2025 (World)*
Coal	1.08 Quads	1.44	5.12	7.48
Crude oil	19.7	29.2	76.9	119
Natural gas	23.5	34.9	88.7	176
Nuclear	0.75	0.81	2.43	2.74
Renewable	6.4	8.9	32.8	50.0
Total	119	171	400	640

Looking at Current Availability and Cost

Of course, availability and cost of fossil fuels can become a huge issue for a country, depending on what a country has in its own reserves. Increased consumption of fossil fuels that a country doesn't have sufficient reserves of leads to increased imports, making that country dependent on foreign nations. The next sections give you an overview of the world's energy reserves as well as how the availability and cost of fossil fuels can make a global impact.

Global availability of fossil fuels

The known fossil fuel reserves worldwide are increasing each year due to advancing exploration technologies. Table 4-6 lists the reserves of various sources (in units of energy so as to show the relative amounts).

Table 4-6	The World's Energy Reserves
Energy	*Reserves, in Units of Energy*
Coal	39,000
Oil	18,900

(continued)

Table 4-6 (continued)

Energy	Reserves, in Units of Energy
Gas*	15,700
Liquefied gas	2,300
Shale	16,000
Uranium 235 (nuclear)	2,600

*There are also around 3,000 EJ of dry natural gas that is currently too far from pipe lines to be economically transported

The total coal reserves are about equal to all the reserves for petroleum products put together. This is either a blessing (more coal, more coal production) or a curse (more coal, more pollution) for the U.S., which has tremendous coal reserves, depending on how you look at it. While domestic coal relieves us of the need for foreign supplies, it's a very dirty energy source. The fact is, American electrical power will continue to rely on coal solely because it's so readily available.

The U.S. and fossil fuels

In the United States, domestic supplies of fossil fuels are dwindling and demand cannot be met at the rate of consumption growth that it's experiencing. Even if new reserves exceed expectations and next-generation oil and gas recovery technologies significantly improve, supply and demand are going to be imbalanced.

Coal is the only energy source that the U.S. doesn't import. In terms of dollars, energy imports accounted for around 24 percent of the U.S.'s $483 billion trade deficit in 2002, and here's how it all broke down:

- 60 percent fossil fuels
- 16 percent natural gas
- 81 percent of the uranium used in nuclear reactors
- 33 percent total net fuel import

The U.S. imports more oil from Canada than any other foreign country, which comes as a big surprise to most people. Next in line is Saudi Arabia, then Venezuela, then Mexico. Each country provides around 1.5 million barrels to U.S. markets every day. The entire Persian Gulf region supplies the U.S. with around 22 percent of its oil imports, or around 12 percent of total U.S. energy consumption. This accounts for a daily cash export of around $1.1 billion.

This trend is not likely to change, as the biggest reserves are being discovered outside of the United States. The Alaska Arctic National Wildlife Refuge (ANWR) has considerable oil reserves, but it is estimated that at the peak of drilling it will only be able to produce less than 1 million barrels per day. The total reserves are estimated at 10.3 billion barrels. By contrast, the Persian Gulf has around 672 billion barrels of proven oil reserves. All told, ANWR may be able to provide the U.S. with less than two years of oil at current consumption rates.

The real cost of fuel

It's interesting to look at the real cost of energy worldwide (the real cost is the cost, adjusted for inflation). Fuel prices have risen and fallen but the real cost of energy has consistently decreased due to improvements in equipment and technology for both the production and consumption of energy sources. If anything can explain the robust increase in energy consumption, falling prices are at the center of the argument. Extraction equipment becomes more complex, but results in cheaper prices. Fuel quality is growing constantly and the prices are still coming down, in real terms.

This is going to change, one way or another.

Chapter 5

Burning Up with Conventional Energy Sources

In This Chapter

▶ Understanding combustion processes

▶ Sorting through common combustion sources

▶ Understanding how raw fuel gets converted into useable power

*T*he vast majority of the world's energy economy comes from fossil fuels of one kind or another. The term "alternative" in fact refers to alternatives to the fossil fuel standard. In order to understand alternative energy technologies and how they fit into the current energy situation, you need to have a general understanding of fossil fuel combustion and the machines that are used to produce ordered power from the raw fossil fuels.

If you're only interested in the alternative energy technologies, you can skip this chapter. But if you're interested in understanding how alternatives fit in, and when and where they are candidates, this chapter is a must.

There are a wide range of conventional energy machines and processes. I start with the basics of combusting fossil fuels and then I describe the most common thermal machines for translating combustion into useable power, or energy.

Combustion Processes

Every fossil fuel must be burned, or combusted, in order to extract energy. The only kind of energy that is available through combustion is heat energy. In some cases, the heat itself is the desired result. Just pile some wood on the ground and set it on fire and — voilà! — you have warmth, heat energy at its most basic. Of course, the key is ordered, or controlled, energy, and for that you need machinery that can manipulate the heat. In home heating systems, the machinery is very simple; it merely channels the heat into the room, or area, where it is used. The simplest home heating "machine" is a fireplace.

In most combustion processes, however, the end goal isn't heat, but electrical energy or energy that can be used for transportation. To produce electrical energy, the heat from combustion is used to spin a turbine that creates electricity via generators. To produce energy for transport, the heat from combustion is translated into mechanical energy, or torque and power on a shaft which is connected to a vehicle's wheels through a transmission (gears).

Briefly, a turbine is simply a fan; when air pressure or steam pressure is set against the turbine (fan) blades it causes the shaft to spin. A generator is like an electric motor, except operating in reverse. When a generator's shaft is forced to turn, two output wires provide electrical power (as opposed to the operation of a fan, where electrical power is supplied to two wires and the fan blade spins).

What I refer to as combustion is actually *oxidation,* a reaction between the fuel being combusted and the oxygen in the air. Without oxygen, there is no combustion.

Hydrogen and carbon content in fossil fuels

All fossil fuels are composed of hydrogen and carbon (among other things that don't factor into the combustion/energy production phase), hence the term *hydrocarbons* to refer to these energy sources. The vast proportion of the useable energy comes from hydrogen, whereas the carbon generates the vast majority of the waste.

When carbon burns completely in an oxygen atmosphere (such as the earth's), the product is a lot of carbon dioxide (CO_2), the gas blamed for global warming. Other byproducts of fossil fuel combustion are carbon monoxide (CO), nitrogen oxides (NO_x), sulfur oxides (SO_x), and various particulates like mercury that cause undue environmental harm.

Gasoline, heating oil, and propane are over 80 percent carbon by weight. Natural gas only contains 75 percent carbon, and this is one of the reasons it's favored by environmentalists. It also burns cleaner and more efficiently, with less extraneous byproducts. In general, the less carbon a molecule contains (in relation to the amount of hydrogen), the cleaner the burning process.

Because energy is derived from the hydrogen content of a particular fossil fuel and waste is related to the carbon content, it's of interest to look at the different fossil fuels and their relative proportions of hydrogen and carbon. Fuels with a high carbon-hydrogen rate are limited in how low a pollution may be obtained via combustion. Table 5-1 lists the percentages of carbon and hydrogen in common fuels. Note that, although wood is not a fossil fuel, I include it in this chart because its energy is also derived from combustion processes.

Table 5-1	The Hydrogen and Carbon Content of Common Fuels			
Fuel	*State*	*Atom % carbon*	*Atom % hydrogen*	*C/H ratio*
Wood	Solid	90	10	9.0
Coal	Solid	62	38	1.63
Oil	Liquid	36	64	0.56
Octane	Liquid	31	69	0.44
Methane	Gas	20	80	0.25
Hydrogen	Gas	20	80	0.25

As Table 5-1 shows, hydrogen gas has no carbon. This is why there's so much interest in developing hydrogen fuels, which exhaust only water when combusted. This is clearly the wave of the future, and I get into more details in Chapter 15.

Wood, as Table 5-1 shows, has more carbon than any of the other sources listed, and it produces a lot of carbon dioxide when it burns. Here's the interesting thing, though: Wood left on the forest floor to rot releases just as much CO_2 into the atmosphere due to natural decomposition as gets released if the wood were burned. That's why wood, despite its high carbon content, is relatively benign to the environment because at some point the carbon dioxide is going to be released, burn or rot. Add the fact that wood is renewable, sustainable, and local in production, and the above chart slanders wood more than is merited. (Head to Chapter 14 for more on wood as an alternative energy source.) Coal, on the other hand, does not decompose in the ground and so coal left alone does not naturally increase global warming.

Efficiency limitations of fossil fuels

The driving force of any heat engine (machine used to derive useable power from the heat of combustion) is a temperature differential. According to Carnot's law (refer to Chapter 3), the efficiency of any combustion process is proportional to the difference in temperature between the heat source (the combustion chamber) and the heat sink (the exhaust environment). In other words, you want to burn the fuel at as high a temperature as possible in order to achieve the most efficiency.

You can also increase efficiency by lowering the temperature of the exhaust environment as much as possible as well, but in most cases this is not a practical variable. The exhaust ambient is what it is, unless huge amounts of energy are expended in lowering the temperature, and that defeats the purpose.

Common Combustion Sources

Currently, fossil fuels are the most common source of heat energy. It's a biggie in the U.S., as Chapter 4 explains, and in developing countries, over 90 percent of all heat energy is derived from fossil fuel combustion. The following sections provide an in-depth analysis of the most commonly used fossil fuel combustion sources.

The heat source for any type of heat engine doesn't *have* to be fossil fuels. Solar radiation, geothermal steam, geothermal water, and nuclear energy are all possible energy sources. Head to the chapters in Part III for more on these alternatives to fossil fuels.

Coal

Coal is a combustible brown or black solid material formed from plant remains compacted under high pressure over the course of millions of years. Coal, a mineral consisting of a majority of elemental carbon, can be found on every continent in the earth. The United States (23 percent of known reserves), Russia and its satellite states (23 percent), and China (11 percent) have significant known coal deposits.

Coal use and consumption

Coal is used predominantly in the generation of electricity. Around 90 percent of the coal used in North America goes toward producing electricity. A ton of coal contains around 21 million Btus, or the equivalent of around 150 gallons of gasoline. Coal is also used for coking (very high temperature heat treating and melting) in the iron and steel industries. In these latter applications, the pollution exhausts can be very high, and the people who work in the plants are prone to a lot of health problems.

In 2000, North America consumed over one billion tons of coal, or about a quarter of the entire demand. Yet at present, China is the world's largest producer and consumer of coal, and this is why China has bypassed the United States in total greenhouse gas emissions. In China, coal is cheap and abundant, and the Chinese government is focused on raising the quality of life for its citizens despite the environmental impact of being so reliant on coal as an energy source. Coal consumption in China is going to increase dramatically over the next decades. This is a sobering thought, as it implies that however hard the United States might try to curtail its own greenhouse gas emissions, China will likely more than exceed our reductions with their major increases. It's an interesting political question, then, whether the United States should spend a lot of money attempting to curtail greenhouse gases while China's economy grows and prospers on the shoulders of inexpensive, dirty coal energy. Is this fair? Critics respond that we've been overusing resources for so long that it's time we pay the piper, but this argument is a tough sell politically.

Coal quality

Coal comes in a wide range of qualities and specific heat values (the amount of heat that a given weight of coal can output, when combusted). The price of coal closely tracks the quality. Classifications of coal are (from soft to hard) lignite, sub-bituminous, bituminous, and anthracite. These ranks are predicated on carbon content, extraneous matter, heating value, and caking or *agglomerating,* properties.

The harder coals are more costly than the softer because harder coals have more energy per weight, and the market bears a higher price. However, softer coals are easier to mine, so there is a cost tradeoff. As far as which industries use which types of coals, the combustion machines are set up to optimize a certain type of coal, so it's the machines, not the end uses, that determine which type a given industry uses. Table 5-2 lists the types of coal and the heat values.

Table 5-2	Heat Values of Different Types of Coal
Coal types	*Heat Value (Btu per pound)*
Lignites (brown coal)	6,950
Bituminous (soft coal)	13,200
Anthracite (hard coal)	13,130
Sub-bituminous (define)	9,000

A pound of anthracite is worth around two pounds of lignite, in terms of energy content.

Peat, found on the surface of the earth, is soft material found in some regions. Basically it's coal in the first stages of evolution, and has relatively low energy density. But since it's very easy to recover it's used widely in some parts of the world for fuel.

Coal combustion

Coal is very high in carbon content, and is very dirty when combusted, as a result. Coal combustion not only exhausts over two billion tons a year of carbon dioxide, but is also responsible for acid rain-forming nitrogen and sulfur compounds. Mercury is also becoming more problematic, with over 200,000 pounds per year going into the atmosphere. To put mercury into perspective, it takes only $\frac{1}{70}$th of a teaspoon to contaminate a 25-acre lake to the point where fish die.

The main problem with coal is that the U.S. has such a tremendous domestic supply, and so the politics and economics of burning coal are politically and economically attractive. Plus we have a sizable domestic reserve, so the investment of capital equipment is low risk, in relation to investing in oil burning equipment.

Crude oil

Crude is pumped from the ground as a semi-liquid called petroleum. It's a mixture of aliphatic hydrocarbons, mainly alkanes of complex combinations of hydrogen and carbon atoms. Oil, like coal, has different grades:

- **Sweet crude:** Oil that's low in sulfur is called "sweet" crude.
- **Sour crude:** Oil that's high in sulfur.

Lower sulfur oil pollutes less than the higher sulfur variety, and costs more in crude form.

Crude consumption

U.S. consumption of crude oil has risen steadily since 1950, and we currently consume around 20 million barrels a day. Worldwide, oil consumption has increased faster than in the U.S. as new markets are opening up at the same time that the U.S. is moving toward more efficiency.

Crude oil products

Various hydrocarbons are classified according to their chain lengths, or the complexity of their molecular structures. The components with longer chain lengths tend to boil at lower temperatures, and so in the refinement process may be individually distilled from the raw crude, resulting in a number of different petroleum products used for various machinery applications.

Crude oil products are classified according to their carbon chain lengths (indicated by the subscript numbers). The products with longer chain lengths generally have lower boiling points, which makes them easier to separate during the refinement phase. Table 5-3 lists the different crude oil products, according to their carbon chain lengths, and tells what common products you'll find them in.

Table 5-3		Crude Oil Products
Product	**Chain Length**	**Use or Common Products**
Gases	C_1 to C_4	Methane is C_4, the most abundant gas
Naphthas	C_5 to C_7	Solvents and thinners — quick drying
Gasoline	C_8 to C_{11}	C_8H_{18} is octane
Kerosene	C_{12} to C_{15}	Also diesel fuel and fuel oil
Lubricants	C_{16} to C_{19}	Very viscous and slippery; used extensively in machinery, and will not vaporize under normal atmospheric conditions, unlike gasoline and solvents
Solids	C_{20} and up	Tar, wax and paraffin, and so on

Note that as the carbon chains increase in complexity, the materials get thicker, or more viscous. The combustion properties also change, in particular the temperature and pressure under which the product burns most efficiently.

In later sections I describe more properties of crude oil products, from the context of the machines that are used to burn them.

Natural gas

Natural gas is a gaseous mixture of hydrocarbon compounds, mostly methane, which is usually found near oil deposits, but sometimes completely independent of crude oil deposits. Methane can also be harvested at rotting landfills and from animal wastes via a process known as anaerobic digestion.

Natural gas is cleaner in the combustion process than heating oil and gasoline, and is used commonly for home heating and cooking applications (as well as hot water heating).

The distribution network for natural gas is cumbersome compared to liquid fuels which can be easily transported via truck and pipelines. So natural gas is most common in urban areas that have the infrastructure in place to provide the gas to a broad range of customers through underground pipelines.

Natural gas consumption

There are over 400,000 natural gas wells in the world, with over 1.1 million miles of pipelines that supply power plants, industry, and homes. In some European countries, natural gas consumption exceeds that of any other energy source.

The U.S. has around 177 trillion cubic feet of proven natural gas reserves (only about 3 percent of the world's total reserves). Natural gas provides around a quarter of all U.S. energy consumption. Only 15 percent is imported, mostly from Canada.

Uses for natural gas

Natural gas is used primarily as a source for residential and commercial heating, as well as large scale power plant electrical production. It can also be used to power vehicles, but this is a minor role because it takes a lot of energy to compress natural gas down to a liquid form, and so the economics are prohibitive. Natural gas also requires special handling equipment and is relatively dangerous. The potential for disastrous explosions is real, and this creates political as well as health problems.

Because natural gas is cleaner than the other hydrocarbon alternatives, most new power plants being built in the U.S. use this fuel. However, clean is a relative term and in the grand scheme of things natural gas is still a high emitter of carbon dioxide. It's a matter of the least evil, with natural gas.

Natural gas combustion

In combustion, methane (CH_4) combines with oxygen to form water and heat and carbon dioxide. Natural gas affords an energy content of around 1,000 Btus per cubic foot, but this depends a great deal on the level of purity. When the gas is impure not only does it burn less efficiently but it emits more pollutants. Impurities also cause equipment maintenance problems, like gummy deposit buildups.

Most natural gas is methane, or CH_4. It's generally found in sedimentary rocks, usually in the presence of crude oil. When it's found with crude, it's called "wet" gas, and when it's unassociated with crude, it's called "dry" gas. Wet gases contain a lot more impurities than dry, and must be refined and processed prior to use.

Natural gas also contains a certain amount of sulfur, which is a very polluting ingredient. The best quality gases, straight from the ground, are low-sulfur, dry gases.

Liquid petroleum gas (LPG)

LPG is similar to natural gas, but it is liquefied, making distribution and use much easier and less expensive. LPG consists of a group of hydrogen rich gases — propane, butane, ethane, ethylene, propylene, butylene, isobutylene, and isobutane. It's derived from a refining process from either crude oil or natural gas. The most commonly available form of LPG is propane.

LPG has an energy density of around 92,000 Btus per gallon, or 2,500 Btus per cubic foot in the gaseous phase. It's used extensively in rural areas where no natural gas pipeline infrastructure exists. Those big, ugly egg shaped tanks located in people's yards contain LPG products of one kind or another.

Over 6.5 billion gallons of LPG were consumed in 2000 in the U.S.: 70 percent went for space heating, 22 percent for heating water, and the remaining percentage went to cooking and appliances.

Historically, the price of LPG has varied quite a bit. It's not uncommon for prices to fluctuate over 20 percent from one season to the next, and this creates hardships for homeowners on fixed budgets because they simply don't know what to expect from year to year.

Home heating oil

Home heating oils include fuel oil, kerosene, and occasionally diesel motor fuel. Most equipment can burn any of these three types with equal efficiency, and so versatility is one key advantage.

About 6.1 billion gallons of fuel oil and kerosene were used in the U.S. in 2000, 84 percent for space heating and the rest for water heating. Over 80 percent of this quantity was used in the Northeastern U.S. and the Mid Atlantic States.

In extremely cold weather, home heating oils tend to gel up, which causes problems with the combusting equipment. Many tanks are located in basements where the temperatures are isolated from the harsh outdoor weather. Adding kerosene helps with this problem, and is a common solution in northern climates. However, kerosene is lighter in weight and has less energy content than the other forms of heating oil, so this solution is economically inefficient (because a tank must be filled up more often), although it may be necessary.

Fuel oils have an energy content of around 135,000 to 140,000 Btus per gallon and are generally considered very energy intensive. For a storage tank of a given size, more energy is available with fuel oils than any other type of liquid fossil fuel. This is advantageous when size is an issue.

Wood

Wood is a nonfossil, renewable combustion fuel. While wood fires are historically very smoky, and release a lot of particulates into the atmosphere, new burning equipment is much better at converting the potential energy into useable energy. When burned the right way, wood is much better for the environment than fossil fuel.

Modern wood-burn chambers are air tight, and carefully control the amount of oxygen used in the combustion process. Temperature can also be carefully controlled so that the absolute maximum of energy is derived from the burn process. In addition, by carefully controlling the parameters, pollution levels can be minimized. Unfortunately, most wood burning is not done optimally simply because the type of people who use wood stoves don't know any better.

Wood creates more ash and mess than other forms of combustion fuels. This may or may not be a problem, but wood-burning homes are generally dustier and dirtier than homes that use petroleum fuels. Wood also causes soot grunge in the air, and creosote buildup in the vent and chimney system.

Wood stoves need constant upkeep, and there's a lot of work in hauling and stacking wood. Plus wood attracts spiders and rodents, which may or may not be a problem depending on the squeamishness of the stove operator.

A cord of wood (4 feet × 4 feet × 8 feet, by definition) weighs about 2 tons and releases around 20 million Btus when burned. This is equivalent to around 145 gallons of gasoline, or a ton of coal.

Scientists have determined that every gallon of gasoline that is combusted was produced by 98 tons of ancient decayed biomass. That's an incredible ratio, and attests to Mother Nature's patience and mankind's greed.

Processing Raw Fuel into Useable Power

From transport to heating to generating electricity, fossil fuels play a major role in our economy. For raw energy to be used for each of these purposes, however, requires that the energy be converted into a useable form — a feat that's accomplished through machinery. The type of machinery used is specific to either the purpose (transport versus heating, for example) or to the type of fuel: a natural gas furnace, for example, uses different mechanisms than does a furnace powered by heating oil.

The following sections look at a variety of different types of machines that are used to produce useable energy from thermal combustion processes.

Converting energy for heating

The most common types of centralized combustion heating systems transfer thermal energy by means of circulating hot air, steam, or water. The most commonly used fuels are oils and gases (including propane).

Heating oil furnaces

High efficiency units have been developed that can use any number of different types of fuels to equal advantage. Heating oil is the most commonly used form of liquid petroleum for home and business heating.

In the furnace, or combustion chamber, of a heating oil unit, the fuel is atomized (similar to the way that gasoline is mixed with air in the carburetor of an auto engine). The fuel-air mixture is ignited by either an electrical spark discharge device, or a pilot light which is on all the time.

The particular advantages of oil-heating machinery are:

✔ There are no pipeline requirements to a central supplier. The liquid fuel is contained on site, in a suitable tank. This is common for homes and buildings located in rural areas.

✔ Oil is a relatively safe fuel. A leak might produce a mess, but the probability of an explosion, or inadvertent ignition, is low, especially compared to natural gas and propane. Consequently, transporting heating oil is less expensive because the fail-safe devices are less onerous.

✔ When a system malfunctions, smoke is heavy and very discernible, making the system even safer because you can easily see when there's a problem.

✔ Oil is a high energy density fuel, which means that the tanks can be smaller for the same amount of potential energy and it's easier to transport than other fossil fuels, and cheaper to boot.

✔ Small systems for camping and remote applications are practical because of the high energy density and safety factor. Most camping stoves and portable heaters use some form of heating oil.

✔ Heating oil can easily be mixed with biofuels such as surplus vegetable oil and cooking grease in order to adjust the burn cycle to make mixtures energy efficient.

The disadvantages of oil heating are:

✔ When the tank goes empty, there's no heat. However, most suppliers of heating oil will periodically check the tank for the consumer, and running out of heating oil is then rare.

✔ The price of heating oil is directly related to the price of crude oil, which means economic shocks can be burdensome. Shocks in the market supply can be predicated on such things as hurricanes, political turmoil, etc.

✔ Many heating systems require electricity, which means that when the grid's down, the system won't work. Some units include backup batteries to overcome this problem.

Methane (natural gas) heating

Natural gas is the most popular energy source for heating in the United States. Supplies are good, and the supply is domestic so interruptions are rare. In urban areas, where pipelines feed the gas directly into homes and businesses, natural gas is especially popular.

A typical heating system includes a thermostat for controlling temperature. A combustion air blower combines oxygen with the natural gas and a spark plug or pilot light ignites the mixture. Efficiency depends on fine-tuning the combustion process, and many natural gas heating machines are poorly tuned and in need of maintenance because there's no way of telling how inefficient a machine is running, aside from unduly high utility bills. A number of safety devices are included to ensure that the explosive gas-oxygen mixture is igniting as it should.

The pros of heating with natural gas are:

- These systems are very efficient in terms of the useable heat energy they provide for the amount of raw energy they consume.

- Natural gas is low in pollution compared to other fossil fuels.

- Exhaust gases are easily vented to the outside world, eliminating the need for a chimney and the associated costs. When a high-efficiency natural gas heating system replaces a conventional oil burning heater in a home or business, the need for extensive venting disappears.

- Natural gas is readily available in most cities and urban areas.

- Natural gas heaters produce very little smoke and soot, as long as they're tuned properly. There's less stink, as well.

The cons of natural gas heaters are:

- Methane leaks are dangerous and may cause explosions, so the systems are inherently less safe than oil burning systems. Most of the time, natural gas, an inherently odorless substance, is given an artificial scent so that leaks can be quickly detected. The scent is similar to rotten eggs, so it's distinctive and difficult to ignore. There are, however, some people who can't smell this additive.

- Price volatility is perhaps the worst of all fuel sources. It's not uncommon for natural gas prices to fluctuate over 20 percent from year to year.

- In rural areas, pipelines are not available, and natural gas is not easy to store on site. Tanks are expensive, and suppliers may be difficult to find.

- Most systems require electricity to operate. When the grid goes down, there's no heat.

Propane heating

Propane is a byproduct in the process of extracting methane from natural gas, or as a byproduct of refining crude oil. Most of the time propane is stored, under pressure, in liquid form. The tanks of gas used by campers and barbecues contain propane. When the liquid is released from the pressurized tanks, a combustible gas results. Propane is odorless, and like natural gas a smelly compound is added so that leaks may be readily detected.

Advantages of propane heating are:

- ✔ It can be used nearly anywhere, since transport is so easy and efficient. Hardware stores and gas stations often sell propane, which is transferred from a big on-site storage tank, into the small 5 or 10 gallon tanks made just for the purpose.

- ✔ Propane supplies for home heating are readily available from utility companies that come out to pump it into on-site storage tanks. Most rural areas have a number of suppliers that do nothing but provide propane for their customers. In the summer, when people don't need fuel for heating, these businesses often do nothing at all.

- ✔ Propane is not water soluble, so there is little risk of ground contamination if it leaks from the tank. Propane heating systems are also less polluting than wood and coal-burning systems.

- ✔ Off-grid homes generally use propane as a primary energy source. You can use propane to power a refrigerator, stove, water heater, and central heater. You can also buy backup generators that use propane, in order to generate electricity.

The disadvantages of propane include:

- ✔ The price per unit output of energy is generally higher than for most other forms of energy. Convenience is the big factor with propane, not cost.

- ✔ Since the energy density of propane is low, tanks are larger and it costs more to transport propane to a tank site. It's also used up faster.

- ✔ If the temperature of propane gets too low (–29 degrees Fahrenheit) the liquid in the tank may not turn into a gas properly, and combustion is a mess, and may not even occur. This is a very low temperature, but in some areas it's not impossible.

- ✔ If tanks are allowed to go completely empty, the interior walls will corrode and oxidize. Odor fade occurs, and this increases the danger level.

- ✔ Many home heating systems require electricity to run.

Powering engines for transportation

Fossil fuels are by far the most common type of fuel used in transport, and there are a range of different types of engines that convert the fuel into useable energy.

Gasoline engines

Most transport vehicles, including cars, trucks and aircraft, use internal-combustion engines that run on gasoline.

In a gasoline engine, gasoline is mixed with air in a carburetor which forms a fine mist (very explosive). The mist is injected into an enclosed cylinder that contains a moving piston confined very tightly into the cylinder. A spark plug ignites the mixture at exactly the right moment, and the piston undergoes great pressure. The spark and the ensuing explosion are referred to as ignition (it's actually a small explosion, to be precise). The piston is connected to a rotating shaft (the crankshaft) by rods and bearings and due to the construction of the rod and crankshaft, torque rotates the crankshaft and provides motive power. The burned gas is exhausted through appropriate ports that open and close at the right phases in the cycle.

In most engines, there are a number of cylinders and the mechanical configuration of the engine provides synchronous, highly tuned operation which provides considerable torque and power to the crankshaft. The crankshaft can be connected to any number of mechanical means including:

- Gear box for an automobile
- Electrical generator for producing electrical power
- Airplane propeller

Gasoline is produced in the refinement process of crude oil. In fact, a majority of the crude oil that is refined ends up as gasoline. Gasoline also comes in grades, with the highest grade of gas containing the highest amounts of octane.

Ethanol is being added to gasoline in many parts of the country, and this helps to decrease the U.S. reliance on foreign oil because ethanol is a biofuel locally produced from corn (and other biomass) products. Ethanol is added at around a 10 percent to 15 percent ratio, and has no noticeable effect on the fuel's performance. In addition to the natural hydrocarbons of gasoline, additives are mixed in to improve the burn efficiencies.

Advantages of gasoline-powered engines include:

- Gas has a very high energy content per unit weight. A smaller gas tank may be used than would be required of other fuels, meaning less weight and size.
- Gas engines are very powerful, given their size and weight. There is no other fuel which affords such high performance ratios. If there were, race cars would use them.

✔ Gas engines can operate optimally over a wide range of temperatures and conditions.

✔ The existing infrastructure for gasoline engines is tremendous. You can buy one anywhere, and get one fixed anywhere. Millions of mechanics are well versed in their operation. Many young men know more about internal combustion than they do about U.S. history.

Disadvantages of gasoline engines are:

✔ Combustion of gasoline produces far too much carbon for a variety of reasons:

   • There are far too many gasoline engines in the world, and cheap ones abound, which are filth spewers of the first order.

   • Gas engines need to be tuned properly or they'll release a lot of carbon monoxide and other poisons, plus they won't run efficiently.

✔ Gas is dangerous. It can ignite very easily, and will explode under a range of real world conditions. Gas is an extremely volatile liquid and changes state from liquid to vapor very quickly under most atmospheric conditions. As the ambient temperature rises, the vaporization increases, which causes safety concerns in hot climates. Different grades of gas are sold in different parts of the country, for this very reason.

✔ Sulfur compounds in gasoline contribute to acid rain.

✔ Because internal combustion engines rely on explosions (albeit tiny ones), they cause a lot of noise pollution.

✔ Gasoline prices can go all over the map, at the drop of a hat. There is so much gasoline used in the world that supply issues inevitably arise.

✔ Gasoline additives (substances added to the basic fuel in order to improve performance or reduce pollution) have been shown to cause cancer in humans. Gas is everywhere, all the time, so this is a real problem. These additives are emitted as exhaust, and humans end up breathing them, or they settle into groundwater and humans drink them.

Diesel engines

A diesel engine is different than a gasoline engine in the way the fuel is ignited. A diesel engine doesn't need a spark plug. As the piston moves up, compression of the gases in the cylinder occur and this causes the temperature to rise. The heat of compression causes the injected diesel-oxygen mixture (from a carburetor) to ignite at just the right moment, and the cylinder is pushed back down with great force. The timing is important, and because the timing mechanism is intrinsic to the engine structure it's more difficult to tune a diesel engine than a gasoline engine. As with gasoline engines, diesel engines are usually multi-cylinder as a way to increase the power output and torque capacity.

Diesel fuel is derived during the crude oil cracking process, similar to the way gasoline is produced. It is used in large trucks and other heavy machinery because diesel engines provide more torque on a crankshaft, and this is necessary in heavy duty applications.

In consistency and combustion properties, diesel resembles heating oil more than gasoline. Vegetable oils and fatty substances can be added to diesel fuel (to a point) without suffering performance compromises. Biodiesel is much more common than bio-gasoline.

The pros of diesel engines include:

✔ Diesel is more energy dense than gasoline, meaning a tank holding diesel fuel contains more energy than does the same-sized tank filled with gasoline — a benefit that's negated by diesel's heavier weight. The higher prices of diesel fuels are offset by the fact that smaller volumes contain more potential energy.

✔ In some locations, and in some markets, diesel fuel is cheaper than gasoline, especially when the higher energy density is taken into account.

✔ Diesel can deliver more power, given the weight of the engine, than gasoline.

✔ Diesel combustion is inherently more efficient than gasoline, so better MPG ratings are the norm for diesel powered cars and trucks.

✔ Diesel combustion produces less CO exhaust than gasoline production.

✔ Diesel engines are more reliable than gasoline because they are inherently simpler. The lifetime of diesel engines is also better.

The disadvantages of diesel engines are:

✔ Diesel contains more sulfur than gasoline does, which exacerbates acid rain.

✔ Diesel exhaust stinks to high heaven, and there's a lot of soot to boot. You can usually tell when a car has a diesel engine because there's a telltale black sooty crud buildup above the exhaust pipe. And the exhaust is often very visible; we've all seen the black smoke belching from a large truck's exhaust pipe.

✔ Diesel engines are finicky, and sometimes they just plain don't want to start. You have to talk nice to them, and give them flowers on Diesel Engine Day or they'll just get worse.

✔ In cold weather, diesel fuel becomes gelatinous and may even crystallize. This can cause catastrophic failure, and it may be impossible to start a diesel engine in really cold weather. The solution is to warm the engine, which is not real easy under most conditions. Starting a fire beneath the hood of your car is not generally advisable.

✔ Diesel fuels are carcinogenic. The exhaust from a diesel engine enters the air and humans breathe the vapors, and the exhaust products also settle onto the ground, and enter the water supply so that humans ingest trace amounts that aren't filtered adequately by the municipal water companies.

Jet propulsion

In a jet engine, air enters the intake and is compressed by the compressor. The highly pressurized air moves back through the engine cylinder to the combustion chamber where fuel is injected, and the fuel-oxygen mixture is ignited or burned. The tremendous increase in temperature of the gas at that point causes extreme pressure in the chamber, and the pressure is channeled through the exhaust nozzle at extremely high velocity. The exhausted gas velocity is much greater than the forward speed of the aircraft, so thrust occurs, and lots of it. The turbine in the gas stream is used to provide power for spinning the compressor, once the engine is running. It also provides power for onboard functions like electronics and lighting, although many jets have onboard power generators that serve to provide this function exclusively.

Jet engines work best at high aircraft speeds (higher than 400 mph) because they are capable of accelerating a small amount of air by such a huge factor. The efficiency of a jet engine goes up with airspeed, as a consequence.

Jet engines require highly refined fuels. Most jet fuel is derived from kerosene, which is obtained during the crude oil refining process, but it can also be made from coal, although this is a more energy intense endeavor, and so less efficient. When kerosene is refined for jet fuel, sulfur content is minimized, as well as the natural corrosive properties inherent to kerosene (you wouldn't want your jet engine to corrode while you're in the air, would you?). The most common kerosene based jet fuel in use is called JET A, which freezes at −40 degrees Fahrenheit. Higher altitude jets must use JET B, which freezes at lower temperatures but is more volatile, and expensive.

To keep from becoming too viscous at cold temperatures, jet fuels contain antioxidants. They also contain substances that keep the fuel from igniting due to electrical sparks (this sounds like a very good idea for anybody who flies on jets).

There is one big advantage of jet engines: There is simply no other energy-producing means that can propel a huge airliner so fast, so far, with so little fuel. In other words, there are no alternatives aside from Superman. Propeller aircraft are powered by gas piston engines, which work better in some conditions. But for simply producing raw power, nothing comes close to a jet engine.

Of course there are disadvantages with jet engines:

- A lot of CO_2 is produced in the combustion process. Jets pollute more per passenger mile than any other type of transport, and the pollution comes out at high altitudes, which creates unique environmental problems. Sulfur content produces acid rain.

- Fuel is very hazardous, and the danger of explosions and fires is very real. The overhead rate for ensuring safety is expensive.

- Noise is extreme. When jet engines are running full bore, the sound can blow an eardrum. And jet engines need to be running full bore in order to maximize efficiency.

- High altitude jets produce contrails, which are white trails of crystallized water vapor, and this affects the natural ambient of the upper atmosphere by concentrating heat into small regions, which end up dissipating and affecting the way the upper atmosphere influences weather patterns.

- Jet fuels are carcinogenic.

- Jet fuel costs rise and fall faster than any other fuel type. The refining process is more involved, and requires high grade raw ingredients which are not always available in good supply.

Utility scale power plants

At the heart of every thermal power plant is the combustion of fossil fuel energy sources. Coal is by far the most prevalent type of power plant in the United States, followed by natural gas.

In all cases (including nuclear), heat is generated and that is used to boil water (or other liquid), which in turn is used to rotate a turbine/generator. A turbine is simply a fan; when air pressure or steam pressure is set against the fan blades it causes the shaft to spin (this is intuitive). A generator is like an electric motor, except operating in reverse. When a generator's shaft is forced to turn, two output wires provide electrical power (as opposed to the operation of a fan, where electrical power is supplied to two wires and the fan blade spins).

Coal-fired power plants

Modern coal-fired plants are very sophisticated, and in the U.S. the pollution control equipment has reduced noxious emissions by a large factor over older type systems.

Coal is mined, then cleaned and de-gassed, then transported to the power plant site via railroad. (All coal plants are located next to railroad tracks because there simply is no other way to transport the massive amounts of coal needed for large scale utility production of power.) Coal is loaded into the hopper which feeds the raw fuel through a pulverizer that turns the hard chunks into a fine dust. A blower forces this dust into the combustion chamber (along with the prescribed amount of oxygen) where it burns at a carefully controlled temperature and pressure. The resulting heat converts liquid water into a very high energy steam, which in turn spins a large turbine connected to an electrical generator. The steam is condensed back into liquid water form, which re-circulates through the system via a feed pump.

Even with the technical advances in pollution mitigation, coal is still a very dirty business, and since we use so much of it, coal fired power plants are a major contributor to greenhouse gases. This is especially true in the developing world, where the pollution control equipment is often substandard, if it's installed at all.

The advantages of coal-fired power plants are:

- Coal is the most abundant energy source in the United States. There is no shortage of coal, and foreign political turmoil does not interrupt the flow of coal.

- Modern plants are very efficient in terms of producing copious amounts of electrical power from the raw fuel.

- Power plants are prevalent, and the technology is mature. This makes for less risky capital investments.

- Capital expenditures for coal-fired power plants are smaller than for other options. In fact, if a plant is built with minimal pollution control equipment, the cost can be extremely low. This is why third world countries build so many coal-fired plants.

- The combustion chambers can use a wide range of different qualities of coal, and this makes the supply issue even more advantageous.

Of course there are a number of cons to coal-fired plants as well:

- Coal combustion is the dirtiest of all fossil fuel energy sources. CO_2 exhaust is very high, as well as CO and sulfur compounds. Emission controls have brought the pollution levels down considerably, but these are costly.

- Coal mining is very harsh on the environment. Some mines completely strip hillsides, leaving ugly, permanent scars. Toxic chemicals are released into the environment near coal mines. These include mercury, lead, and arsenic. Ground water pollution is a real problem near coal mines.

- ✔ Coal mines are dangerous and the labor is hard and unhealthy.

- ✔ The cost of transporting coal is high, and the cost of machinery to move the massive amounts of stinky, tarry material is also high. Coal trains are noisy and environmentally invasive.

- ✔ There is a waste problem associated with all the ash and soot leftover from coal combustion.

- ✔ The water used to make steam becomes polluted after a time, and this is a disposal nightmare.

Methane-fired power plants

A well designed methane-fired power plant is more efficient than either coal or oil fired plants, plus they pollute much less. Note that there are two electrical outputs. Instead of wasting the heat from the combustion cycle, it is channeled into a second boiler/turbine system. This enhances the efficiency of the process by a large margin.

Advantages abound:

- ✔ Methane affords low carbon dioxide, carbon monoxide, nitrogen oxides, and sulfur oxides pollution levels over coal and oil.

- ✔ These power plants are very efficient in terms of their power output divided by the amount of raw fuel they consume.

- ✔ Methane is readily available around urban areas where large power plants are used the most.

- ✔ Pipelines supply the raw fuel, which obviates trains and other ground transportation.

- ✔ Methane plants can be modified to burn hydrogen gas (there is no carbon pollution at all with hydrogen gas). While hydrogen gas is not yet available, investing in methane-fired power plant infrastructure is wise because when hydrogen does become available, the equipment will largely be in place (modifications are needed, but these are much less costly than starting from scratch).

And of course, in the far corner are all the disadvantages:

- ✔ Methane is dangerous stuff. If it leaks into the air, it can explode in spectacular fashion.

- ✔ Price spikes are common, so the cost to the utilities varies and this gets passed on down to the customers.

- ✔ Methane exploration and recovery is environmentally damaging.

- ✔ Methane does produce carbon dioxide, albeit in smaller amounts than coal or oil, but it's still there.

Oil-fired power plants

Relatively rare are oil-fired power plants. They operate basically the same as a coal-fired plant, but with a large oil combustion turbine in place of the coal furnace (a combustion turbine resembles a jet engine). There are very few of these being built now, since methane-powered plants have taken over as the generator of choice, when it comes to fossil fuels.

Oil is safer than methane, and the supply lines for oil are more common than those of methane (which require pipelines). The combustion turbines can be modified to accept coal or methane, but this is also rare. Oil is a high density fuel, so the entire process is more efficient, weight-wise, than a coal plant.

But the pollution levels are high, and the overall efficiency is nowhere near as good as a combined cycle system.

Chapter 6

The First Alternative-Energy Sources: Efficiency and Conservation

The U.S. can save up to 10 percent of the energy it uses by simply using energy more efficiently. Plus, another 10 percent can be saved by investing only 5 percent of the current U.S. energy budget in new technology and infrastructure.

The point? That even without resorting to alternative energy technologies (a whole bunch of which are explained in Parts III and IV), mankind can do much more to use current conventional resources more efficiently. Major increases in efficiency are coming down the technological pipeline all the time. Cars are growing more efficient, as well as appliances and lighting systems. Building codes are mandating higher efficiency heating and cooling systems. And perhaps most important of all, hybrid and electric auto technologies will radically improve the efficiency performance of the transport sector. Because of these advances, the first alternative energy source is using energy more efficiently and practicing conservation.

Most books on the subject don't take this approach, but in this chapter I explain how this could work. I describe how efficient technologies have affected our energy markets, how big players — utility companies, communities, and so on — are getting in on the efficiency act, and how each of us can make a difference. If enough people begin practicing efficiency and conservation, the U.S. could improve our situation drastically, without even investing in expensive and risky alternative energy propositions.

Efficiency, Consumption, and the Energy Market

Efficiency is not the main factor in reducing energy consumption. While it seems only logical that rising efficiencies will result in reduction in demand, the truth seems to be that increasing efficiency actually increases demand.

So if efficiency isn't the main factor, what is? Price. The only way that consumption will ever be reduced is by raising the price of energy. Because efficiency gains result in increased energy consumption, it's up to the government to tax consumption to the point where people use less energy. This is a difficult proposition, because Americans have become used to low energy prices. Plus, it will take bravery on the part of politicians to push such policies; there doesn't appear to be much of that commodity on hand in today's politicians.

In the following sections, I walk you through the different pieces that make up the puzzle of the energy market, and I give you enough basic info to demystify the factors that impact the market and the ability of society to be more efficient with its energy use.

Efficiency increases demand

After the Arab oil embargo of the '70s, the governments of the world acknowledged a need to consume less energy, and a number of policies were instituted to effect fundamental changes. But America's total energy consumption has risen by over 30 percent since the oil embargo of the 1970s and electrical consumption has risen by 50 percent.

Per unit of energy, the U.S. produces twice as much economic output as it did in 1950. This is an amazing increase in efficiency, but during the same time period, U.S. consumption of energy has increased threefold. Figure 6-1 shows the efficiency of the U.S. economy versus total energy consumption.

Figure 6-1:
Energy cost of the U.S. economy versus total consumption.

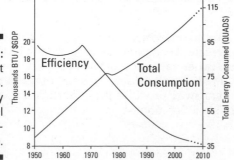

Carbohydrates versus hydrocarbons

Some people wistfully look back on the days when the economy was powered by carbohydrates, as opposed to hydrocarbons. One such group, the Institute for Local Self-Reliance, exhorts people to revert to the "carbohydrate" economy, or agriculture. They promote farming — specifically a pre-mechanized agricultural society — as the solution to the current energy woes. At first glance, you may think that that's not such a bad idea, but here are some facts:

✔ Drilling for oil takes very little land area compared to growing carbohydrate energy sources — about 1,000 times less land area, in fact, on a per unit of energy basis. If you're concerned with deforestation, carbohydrates are not the solution because forests are cleared and land is tilled to make way for productive agriculture.

✔ A car engine, including all the refineries and distribution networks that make it work, is 16 times more efficient at converting crude oil to locomotion than a meat-eating cyclist who relies on a vast grid of farming, cultivating, livestock production, slaughter houses, distribution systems, grocery shelves, transport to and from the grocery store, and finally cooking and eating and digesting in front of the television.

✔ Plowing land for growing carbohydrates adds more carbon to the air than mining for coal or drilling for oil because farms require such huge amounts of cleared land, and the machinery used on farms is very energy intensive.

✔ Economies in the parts of the world that rely on carbohydrates are always far behind economies that rely on hydrocarbons in terms of productivity and quality of life. This is not simply a reflection that these "backwards" economies don't have our sophisticated machines, it's a fact of energy production. Thriving economies' energy production is much more efficient at producing and using useable, or ordered, energy. The production is more concentrated, and this is good for the environment. It's not the production of fossil fuels that harms the environment, it's the byproducts of using those fossil fuels that is becoming troublesome.

In short, the carbohydrate economy is much worse for the atmosphere, in terms of pollution, than the current hydrocarbon economy. This illustrates a key point you must remember: The way things appear to be are often not necessarily what they are. Living creatures, especially farm animals and horses, require a lot of land, and they're not efficient at converting that land into useable energy. Machines are efficient and powerful, and they produce high-grade, refined energy that performs magical tricks.

American economic output per unit of energy is the highest in the world. Americans provide by far the greatest economic stimulus to the world economy, and everybody benefits. Higher efficiency allows people to do more, and do it in less time. Time is the hidden variable, and due to mankind's intrinsic impatience it's time that swamps all the other variables. Time is money, and for most people it's a lot of money.

The 55 mph mandate is a case in point. Following the Arab oil embargo of the 1970s, the U.S. government passed a federal speed limit of 55 mph. While it's true that driving your car at 55 mph saves gasoline, nobody paid much attention. The reason? It took longer to get to where you're going. Mandating slower speeds comes with a real cost, that of man hours. The fact that people place more of a premium on their personal time than on the amount of gas mileage they achieve gets to the very foundation of the problem with energy: People don't just look at the total amount of energy they use; instead, they look at convenience, time, energy-consuming processes, and a whole host of other factors (refer to Chapter 3). Figure 6-2 shows how the efficiency of energy consumption per passenger mile has improved while our total aggregate consumption of energy has risen.

Figure 6-2: Efficiency of transportation versus total U.S. consumption.

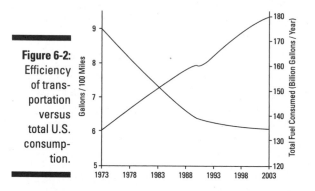

Increasing population, increasing demand, and a bit of bad advice

In the early 1980s, efficiency advocates assured everyone that the demand for electrical power would remain flat well into the future. Their reasoning held that increases in efficiency would offset the increase in demand for energy. Things didn't quite turn out that way. In fact, demand has doubled, in spite of huge advances in efficiency and conservation measures. Part of the problem is that the world's population is constantly increasing, and even if each individual uses less power, there are more people requiring it.

Not only were the prognosticators wrong, but their assumption that demand would remain flat has led to some terrible problems. In California, for example, no new power plants have been built in over 20 years. As a result, California has experienced peak-energy supply problems that resulted in Governor Gray Davis being booted out of office (of course, the fact that his opponent was Arnold Schwarzenegger, a famous movie star, didn't help matters, voters being what they are, especially in loosey-goosey California).

If you simply plug in a more efficient light bulb, like a fluorescent, a net gain in efficiency results. If you put a smaller, more fuel-efficient engine into an auto, the result is an efficiency gain. But that's not how things work. Most people don't typically replace working parts with new parts just to gain a bit of efficiency. People don't plug in a new light bulb until the old one burns out. People don't replace an old engine with a new one; they buy a whole new car. Basically, people replace old systems with new systems, and then still use the old systems. The net effect is to increase the amount of systems and, as a result, the total aggregate amount of energy consumed.

In order to reduce energy consumption, a new, more efficient technology (of which there is no shortage) must have a bigger impact in the new markets that are created than the old markets that are displaced. New efficiency technologies must displace the old technologies, and when these new products enter the market, new uses for those products arise so that new markets swamp the old markets.

Lighting is a case in point. There are millions of new uses for LEDs, which are an order of magnitude more efficient than the incandescent bulbs that they replace. Jumbotrons are everywhere now: in advertising signs, at ballparks, along roadways to direct traffic, and so on. While an LED light can be plugged into an existing socket in a home and achieve better energy efficiency in the home, the myriad of new uses for LEDs has dwarfed any efficiency gains that have resulted from replacing existing lighting schemes.

Price increases and decreased energy use

The real price of energy has consistently gone down over the last 100 years (I explain why in Chapter 3). As the price of energy has fallen steadily, consumption has risen. So it makes sense that, if prices rise, consumption falls. And this is, in fact, what happens.

In 1979, during the Arab oil embargo, for example, Americans cut back on driving, purchased smaller cars with better mileage, and dramatically reduced their use of oil. The motivator? The price spikes that accompanied the embargo.

In the four years after 1979, the country weaned itself off of over 3 million barrels per day. Not until 1997 did the country get back to the same level as 1978. But half of that cut was in residential fuel usage for generating electricity. Americans switched a lot of power generators over to coal and natural gas. The U.S. didn't really reduce its use of energy as much as it might seem, given the record — the country simply played a shell game and switched to different sources. While this was good for the economy, and for political independence from foreign sources, it didn't make a lot of difference to the environment.

Aware of the connection between prices and consumption, federal and state governments increase taxes on certain types of energy sources specifically to impact consumption levels. These taxes increase the cost of gasoline by 40 percent and electricity costs 20 to 80 percent. The Clinton Administration proposed a comprehensive tax on energy, but it was never effectively facilitated. In Europe, energy prices are roughly twice what they are in the U.S., and this has resulted in Europeans using much less energy per capita than Americans. So there is no denying the effect of higher energy taxes. But the politics is very problematic. Any politician that advocates higher energy taxes is not a politician for very long.

Back to the future, from 1979 to now

We're facing an energy crisis now. So the big question is, Can we learn to conserve again, as we did during the energy crisis of the 1970s? Things have changed a lot since then:

✔ We're bigger and busier now, and our economy is nearly five times as large now as it was then. On average, we drive nearly twice as many miles per day as we did in 1979.

✔ The current high price of fuel is actually lower, in real terms, than it was in 1979. This means that fuel price fluctuations don't make as big a difference now as they did in 1979. Fuel prices are a smaller percentage of our economy, both personal and as a whole, than they were in 1979.

✔ The 1979 crisis was predicated on the supply side; the Arab countries reduced the supply, despite the fact that demand was as high as ever, and prices soared. Since then, many other regions have begun supplying oil, like Mexico and the North Sea and Alaska. But unlike in 1979, there are no huge new reserves about to open up to drilling.

✔ In 1979, stagflation, unemployment, and a miserable economy prevailed, and consumers simply stopped using so much oil. Today our economy is much less sluggish, even in spite of the credit crunches sweeping the world.

✔ Globally, the U.S. is a smaller piece of the pie than it used to be. While our economy has been growing like crazy, other economies have also been growing, and at a faster pace. In 1979, we used 29 percent of the world's oil, now it's only 24 percent. In all, global demand for oil is expected to rise by over 1 million barrels per day. It's what's happening in China and India that matters now.

✔ Like the boy who cried wolf, economic impacts of energy price surges in the past have taught that things will eventually iron themselves out. It will take a major calamity to get people motivated to conserve the way we did back in the seventies. Back in those days, we believed in the wolf; today we're much more jaded (for better or for worse).

On the flip side of the coin, efficiency gains have become much more permanent, and consumers are clamoring for more fuel efficient cars and trucks. People understand efficiency and conservation much better than at any time in the past. And people are also beginning to understand how important the environmental effects of energy consumption are. Back in 1979, it was all about economics; now it's a two-fold problem — economics and environment.

Big-Picture Strategies for Reducing Energy Consumption

On a large scale, entire communities are making inroads to reduce energy consumption. Governments are mandating green technologies, and systems that will help to reduce energy consumption. Automakers and industries are also improving technologies that afford better efficiencies. The trend toward conservation and efficiency is very powerful and is growing by leaps and bounds each year. I cover the efforts toward conservation in the sections that follow.

Dealing with peak-demand problems

On a macro level, one of the best ways to reduce energy consumption is related to the peak-demand problem. Peak demand occurs when a utility company is called upon to provide more power to their customers than they are capable of providing, given their current resources. As a result, small generators are turned on, and these are much less efficient in several regards (go to Chapter 3 to find out why bigger generators are more efficient).

By requiring all homes and businesses to use TOU (time of use) meters, which charge different rates depending on the time of use of the energy, this problem can be solved. In California, peak-power prices are around three times off-peak prices (31 cents per kWh versus around 11 cents per kWh). This reduces not only aggregate demand, but it also reduces demand during the peak times. It also reduces pollution.

Expect to see more TOU metering schemes in the future because they make a lot of sense in conservation terms. People may not like the fact that they can't use their air-conditioners whenever they want, and as much as they would like, but they do respond to cost differentials.

Dealing with urban sprawl

Urban sprawl fragments wildlife and Mother Nature's natural designs. Low-density housing developments take far more resources per dwelling than high-density city habitats. And the public transportation options are severely limited, meaning most people own cars and use them quite a bit. Infrastructure systems such as water, sewer, trash, and electricity all have to be stretched out, meaning more resources are required to deliver these services to customers.

Cities easily provide more opportunities for energy efficiency. High-rise housing affords no opportunity for landscaping and the water usage that goes along with it. Community housing structures have common walls, so the energy used per resident is lower. Water, sewer, and trash require fewer resources per resident. The list goes on.

The reality, however, is that people are constantly leaving the dense housing of the cities for the more spacious lifestyles of the suburbs. The quality of life is often much better, and the elbow room is essential for certain facets of mental health.

The solution would seem to be to integrate the best of both worlds, and in many suburban towns this is the goal.

Promoting green communities

Different types of communities have different attributes that can make becoming energy-efficient more or less accessible. With the interest in green living on the upsurge, many cities and states have taken steps to make themselves havens for environmentally conscious citizens. But maybe you don't want to live in a community at all. Perhaps you want to join a commune or you're interested in going off-grid, or back to nature. Each situation has its pros and cons, though more cons are associated with going it alone.

The best communities are those that dedicate their fundamental philosophies to going green, and more and more of these are cropping up all the time. Green cities promote environmental policies aimed toward making it easier for their citizens to be kind to Mother Nature. The characteristics outlined in the following sections are indicators of municipalities that take energy efficiency and pollution reduction seriously.

Air quality

In an effort to improve air quality, over 250 cities have committed to conforming to the Climate Protection Agreement, which encourages cities to reduce their greenhouse gas emissions by 7 percent from 1990 levels. Some of the best cities include San Francisco, California; Portland, Oregon; Austin, Texas; Boulder, Colorado; and Cincinnati, Ohio.

A list of the top ten green cities is put out annually by National Geographic's *The Green Guide,* which informs consumers about a wide range of green issues. Check out www.thegreenguide.com.

Public transportation

Commuting is a big factor in finding an energy-efficient community. In general, cities that take green living seriously provide good public transportation, which enables citizens to save a lot of energy on transport costs. Public transportation is also much less polluting than having thousands of cars on the road, so everybody's air quality benefits. You may or may not want to use public transportation, but the fact that it exists in a workable state says a lot about a city's general attitude. Check a city's public transportation by using it — or trying to, as the case may be.

Utility structures

When you check a city's utility structures, take the following into consideration:

- ✔ **Where does the city get its energy?** Is it nuclear? Derived from coal plants? Driven by hydropower or other renewables?

- ✔ **How much pollution per kWh does the city's utility generate?** The average number of pounds of carbon dioxide per kWh is 2. Anything less is good. Any community that gets power from a nuclear reactor will be way below the average.

- ✔ **How much does power cost?** The fact that power is expensive may be a good thing, even though it costs more. Expensive power means that efficiency investments are more cost-effective, and efficient homes are worth more because they're cheaper to live in. And when power rates are high, the community is likely to be cleaner because less power is used.

   Check out a community's utility rates by going online to the Web sites of the various utilities. You can check the online yellow pages under "utilities" to find the names — and sometimes even the Web addresses — of a community's utilities. You can also check out www.sustainlane.us and www.eere.energy.gov/greenpower for lists of cities that promote and subsidize the use of alternative energy.

- ✔ **Is a tiered rate structure utilized?** Tiered rate structures allow residents to use as much or as little power as they want, but the tiered rates penalize profligacy.

Recycling programs

Recycling is possible for any household, but it helps if your community has set up a formal regimen whereby recycling is centralized and encouraged. When this is the case, the costs of transportation and processing are borne equally by all citizens through utility bills. The better communities provide recycling containers that you set out on the street with your regular trash. The best provide different containers for different types of recycled materials — for instance, aluminum and metal, plastics, newspapers, and boxes. Some communities even remove lawn clippings and organic refuse.

Water supply

A city's water supply is critically important for a number of reasons. First, it must be pristine and clean. Otherwise, you'll have to buy drinking water or a filter system. Second, some cities simply don't have enough water. You can conserve, but if too many hands are reaching into the pie, costs are likely to rise, as is the amount of pressure to reduce usage even more. Find out whether the community has installed water meters. If so, homeowners are a lot more likely to use less water. In communities where water meters don't exist, the people have absolutely no incentive to use less water.

Housing designs

In green communities, home designs are ultra-efficient, with solar exposures and strategically planted deciduous trees. Homes feature thick insulation, double-pane windows, window coverings, automatic awnings for hot summer days, solar attic vent systems to purge hot air from attic spaces, whole-house fans, and solar light tubes in all kitchens. Skylights are used as much as possible, along with passive solar heating and cooling arrangements.

Some communities specifically demand sustainable housing designs. The U.S. Green Building Council (USGBD) runs a program called Leadership in Energy and Environmental Design (LEED) that sets criteria for what constitutes green building. These criteria include things like:

- ✔ Insulation properties in the walls and ceilings
- ✔ The design and placement of windows in a building (called *fenestration*), which affects not only the insulation properties but also the availability of solar lighting and heating
- ✔ The type and operation of any HVAC systems
- ✔ Use of water resources and on-site means for conserving water
- ✔ Installation of any solar energy systems

Making transportation more efficient

Two-thirds of the oil being consumed is used for transportation, so the most effective and immediate area to look for material improvements in fuel consumption reductions is transport. Improving fuel efficiency and conservation both offer their own gains.

In the U.S., the average fuel efficiency of a vehicle is 55 percent better than in 1960. The ban on leaded gasoline, higher fuel economy standards, and the widespread use of catalytic converters have decreased the pollution levels (per vehicle) by 60 percent. But the number of vehicles on the road has skyrocketed, doubling since 1960. The number of miles driven per year has tripled, from less than 600 billion miles to well over 1.6 trillion miles (wow . . .). Transportation is responsible for the release of half of nitrogen oxides, which cause smog and acid rain. And transport accounts for over 85 percent of carbon monoxide emissions. Each year, America's 128 million autos and trucks account for nearly 300 million tons of carbon being released into the environment.

Because efficiency gains, on their own, only lead to increasing energy consumption, not decreasing, reducing energy consumption requires conservation and efficiency improvements to go hand in hand. Unfortunately, since the oil shocks of the '70s, Congress has done nothing but relax the MPG standards, and the efficient technologies have not grown the way they could have. To make matters worse, tax policies have actually favored big gas pigs, like the Hummer H2.

Making more efficient autos

Cars are mostly bulk: huge masses of metal and steel that do little more than channel power and energy to where it's needed. With the advent of electrical power systems, these bulky subsystems will give way to smaller, lighter, more fuel efficient systems that offer the same quality of performance as the current technologies. Some examples:

- **New power semiconductors** displace thousands of pounds of grinding, smoking, burning mechanical and hydraulic parts with ounces of semiconductors. (Not only are these drastic improvements being realized in autos, but also in large manufacturing plants that use laser welding machines and complex robotics.)

- With **intelligent processors,** there's less pollution. Emissions can be reduced by 50 percent and efficiencies can be raised by more than 20 percent. Less use of fossil fuel lubricants, less use of fossil fuels for drivetrain power. Less use of all sorts of noxious compounds. And intelligent efficiency increases don't reduce pollution in the same way that catalytic converters do, by scrubbing fossil fuel emissions and reducing efficiency in the process. They reduce emissions by simply reducing fuel consumption in the first place. This is a much better way to achieve the same end result.

✔ **Electrical control systems** can run internal combustion vehicles at a much finer degree than the current mechanical control systems (valves, fuel injectors, etc.). Emissions can be further reduced, as a consequence.

Equivalent improvements can be made whenever combustion and thermal processes are used in manufacturing processes to treat materials. There are literally billions of valves and firing systems that now control combustion processes, and when these are outfitted with intelligent electrical controls the improvements in efficiency and emissions are remarkable. In particular, hybrid cars rely on batteries and electrical powertrain systems for their drive power a majority of the time. When the batteries simply can't keep up with demand, a small diesel motor kicks in and runs at a very specific rate of speed, with its load also carefully controlled. This results in much better efficiency and emissions, without compromising the quality of the ride. Head to Chapter 19 for more information about hybrids.

Fuel demand would be much higher today if it weren't for the impressive increases in fuel efficiency of modern autos, and pollution would be much worse without catalytic converters and the many techniques employed to reduce toxic wastes from internal combustion engines. Witness the terrible air in China, which has very few internal combustion engines compared to the U.S. but much worst air quality. Imagine if every Chinese citizen owned a car, and add to that thought the notion that their cars would not have much by way of pollution mitigation equipment. The world needs Chinese cyclists.

Conserving fuel with higher MPG ratings

According to the Sierra Club, the U.S. could easily reduce its oil consumption by over 15 percent, or the equivalent of 3 million barrels a day of oil, by simply increasing the SUV standard from 20.7 MPG to 40 MPG.

Auto manufacturers currently have the technology to build conventional passenger cars and trucks that average 30 to 40 MPG without sacrificing power, performance, or even comfort. A report written by the Union of Concerned Scientists (UCS) includes detailed plans for two different SUV designs equivalent in size to commonly available current SUVs, but vehicle designs in the report average over 27 MPG and 36 MPG. Adding the necessary technologies would only increase the cost by $600 and $2,500, respectively. In each case, thousands of dollars would be saved on fuel costs, thereby recouping the added costs in a matter of a few years. Why haven't these changes been implemented? The market simply doesn't care at this point in time.

CAFE and MPG performance

Gains in fuel efficiency were prodded by the 1973 Arab oil embargo, when Congress enacted the Corporate Average Fuel Economy (CAFE) standards that are still in place today, in varied form. In 1973, cars averaged 13.5 MPG (miles per gallon) and nobody really cared. Trucks, minivans, and SUVs averaged only 11.6 MPG. Congress required manufacturers to increase fuel efficiency to 27.5 MPG for cars and 20.7 MPG for light trucks.

So what happened? In the U.S., the average MPG performance of vehicles fell to its lowest overall record since 1981, predominantly due to increased sales of light trucks and SUVs. So while efficiencies increased, consumers demanded larger vehicles and the efficiencies were more than offset.

In the last two decades, SUV sales have increased 20 fold, making up 25 percent of total automobile purchases. The average vehicle in the small truck/SUV category has a fuel efficiency rating 30 percent lower than autos. The lower fuel economy of light trucks and SUVs results in a 40 percent increase in greenhouse gas emissions, and more than a 50 percent increase in nitrogen oxide. People simply switched to SUVs and trucks, in response to the government regulations.

While improvements in passenger cars have been impressive, albeit behind the mandated goals, buses and heavy trucks are still the most significant polluters on the planet. Single trucks can emit up to 150 times more noxious chemicals into the environment than a passenger auto. Diesel fueled trucks are not required to have pollution control devices, and as a consequence they release prolific clouds of fine dust, soot, and toxins into the air. Air quality studies indicate that trucks cause 125,000 cancer cases and 50,000 premature deaths due to their diesel exhausts.

It is estimated that 90 percent of the pollution could be eliminated from trucks with effective mandates. And fuel efficiency for all transport could be increased by over 30 percent. The problem is that trucks have lifetimes of over 25 years, and the mandates will only apply to new trucks, which will become more expensive as a consequence. And as a consequence of their being more expensive, decisions to buy new trucks to replace the old will be put off longer, thereby increasing the lifetime of old trucks even more. In this case, mandating strict efficiency and pollution standards will likely result in more pollution and fuel consumption, not less. This is a difficult paradox to overcome.

Another study concluded that if the U.S. were to shift its usage of passenger autos to hybrids (like the Toyota Prius), Americans would reduce their oil consumption by 50 percent.

Efficiency and Conservation Changes You Can Make in the Home

Despite the fact that energy efficiency technologies end up creating even more demand for energy, the game is not lost. There are a wide range of things people can do on their own to conserve energy use, and most of these are relatively straightforward, and practical. In this section I describe some of the best ways to save energy, and in the process, reduce pollution.

For a much more complete coverage of energy efficiency in the home, consult my book *Energy Efficient Homes For Dummies* (Wiley).

There are a lot of things you can do to use your current appliances and energy consuming systems more efficiently, as the following sections explain.

Seal and insulate your home

You probably spend more energy on heating and cooling than you do on all the other energy-consuming functions combined. This means the greatest potential for energy savings lies in finding ways to make your heating and cooling efforts more efficient. Sealing and insulating is where you can find the biggest energy-efficiency improvements for the least amount of cash and labor on your part. Most homes have problems that can be fixed for $100 or less — in fact, sometimes all you need to invest are a little time and labor. Most people have not even inspected their homes for leaks, and the most obvious ones are the worst offenders. Just by visually inspecting ductwork and window seals you can usually find the big culprits within half an hour's time. Seal with caulk; it's easy and quick and effective.

Tune up your heating and air-conditioning system

Most homes have both heating and cooling systems. Together, these account for a large portion of your total energy use. (To find out just how much of your energy consumption is devoted to heating and cooling your home, do a home energy audit. Refer to *Energy Efficient Homes For Dummies* for a systematic procedure.) If you haven't had your HVAC serviced in over five years, it'll be worth the $100 cost because almost every system can be tuned to run better and more efficiently.

Use efficient bulbs

The typical North American home spends between 8 and 10 percent of its energy budget on lighting (not including the bulbs and equipment). In any given night, the average home uses 32 light bulbs, including not only lighting for rooms, but lighting inside refrigerators, ovens, and microwaves. Most of these light bulbs are the standard, 50-cent, screw-base style using a technology that has been around for over 100 years. People have spent the vast majority of their lives with Thomas Edison's original brainstorm. But things are changing. A number of choices are available now that, while more expensive up front, provide much better economics and performance over the lifetime of the product. They also offer lower pollution.

Fluorescents led the way, and the LEDs are quickly catching up. Most homes use a lot more lighting than is strictly needed, so take a critical look at how and when you're using lights. You'll find that you can cut it back by around 50 percent without sacrificing much quality of lifestyle.

Reduce water consumption

The minimal human requirement for water is only around four gallons per day, and that's for both consumption and cleaning. So how much does a typical American family (two adults, two kids) use? Between 200 and 350 gallons of water a day. That's 73,000 to 128,000 gallons per year. Landscaping, swimming pools, and spas take even more water. A small, working ranch may use over 270,000 gallons per year. Bluntly put, that's a well of a lot of water.

In most homes, water has four main destinations: faucets, showers (baths), dishwashers, and washing machines. In each case, water heating may be required, which consumes around 20 percent of a typical home's energy costs. Other uses of water include outdoor irrigation, pools, spas, fountains, and so on. By conscientiously rationing your water consumption (especially hot water), you can easily reduce it by up to 50 percent.

Use appliances more sensibly

Including your domestic water heater, around 40 percent of your power bill is consumed by your appliances, big and small. And just because an appliance is small in stature doesn't mean it's small in energy consumption.

Understanding how much power your appliances consume is the first step toward recovering from the rampant disease of appliancism. You can lower your power bills by using your appliances more sensibly — or even better, by not using your appliances at all. For instance, a great way to lower your power bill is to set up a clothesline. My personal experience with this particular subject is illuminating. When I do energy audits, I suggest clotheslines and inevitably get a grimace. "What?! A clothesline?" But the vast majority of people who set up a clothesline and give it a try end up using it habitually, and it isn't a big burden. Clothes smell better when they're dried on a line, that's a fact.

Don't trade in your old appliances for energy-efficient models until the old unit is well past its prime. Simply upgrading to more energy-efficient machines doesn't necessarily help the environment. If you buy a new refrigerator, and throw the old one away, you may be saving raw fuel costs but in the end, the new refrigerator took a lot of invested energy to manufacture and the old one takes a lot of energy to properly discard.

Exploit sunlight

Sunshine is free (at least until the government comes up with a way to tax it). And sunshine is natural and amiable, so it should be used as much as possible. In fact, most people do use sunshine, but not nearly as effectively as they could.

Using sunshine effectively isn't as simple as just letting the sun shine in. Obviously, in the winter, you want both the sunshine and its heat in your home as much as possible. In the summer, you want the sun's light but not the heat — two goals that are almost always at odds.

You can use sunlight to heat your home in the winter and avoid doing so in the summer. You can build sunrooms that increase the value of your home and provide increased living area at an efficient cost.

For much greater detail on this subject, check out *Solar Power Your Home For Dummies* (also authored by yours truly and published by Wiley).

Use ventilation, fans, and air filters

Unless you live in a perfect climate where it's never too hot or too cold, heating and air-conditioning make up the largest component of your power bill. Ultimately, what you're buying is comfort, because you don't actually *need* air-conditioning, and most of the time you don't need heat. If you rely on your heater and your air-conditioner to keep you comfortable, then you're missing a big opportunity for energy efficiency. The key to comfort is air movement.

By moving air appropriately through your home, you can achieve a much higher level of comfort and save money by using your HVAC system less, particularly in the summer.

Chapter 7

Understanding the Demand For Alternatives

* *

In This Chapter

▶ Getting a grasp on global warming and environmental problems

▶ Watching supplies dwindle and demand rise

▶ Evaluating the political impacts

* *

*B*elieve it or not, energy consumption has been advantageous to humanity: Energy has made possible the technological advances that have brought humankind from the primitive to the advanced. We live longer, healthier lives: In the U.S., for example, life expectancy has increased 66 percent over the last century. We are more secure and knowledgeable about our world. Life is just plain better than it used to be, and this has been made possible by advances that took a great deal of energy. To make any kind of claim that energy consumption has been anything but advantageous to humanity completely misses the point. So what is the point? The biggest problem with our current energy is our reliance on fossil fuel energy sources.

In this chapter I review the environmental consequences of relying so heavily on fossil fuel. Global warming may be the biggest problem facing humanity, but acid rain, smog, ground water contamination, and other problems are also evident. There is also the inescapable reality that fossil fuel sources are dwindling. Even if there were no environmental damage from fossil fuel combustion, we're going to run out at some point.

Environmental Primer

In 1969, Congress passed the National Environmental Protection Act, which was signed into law on January 1, 1970. The law was an acknowledgement of the importance of balancing the unimpeded growth of human population with its effects on the environment. For the first time since the Industrial Revolution began, national planning was mandated to begin including the problems inherent in urbanization, resource exploitation, and population explosion — problems that affect our environment in negative ways. For the first time since the Industrial Revolution began, we started looking at the big picture of energy consumption, taking into account the many different facets that come into play.

Expansions of the initial law have been passed consistently since 1970, and environmentalists now have a wide range of tools for altering and influencing the planning process of new power plants and energy schemes. The results have been profound, and major successes have been achieved. For instance, many pollution issues like littering, polluted lakes and streams, and so on, have been addressed. Automobiles are much cleaner now than they were 20 years ago, and some communities that had terrible smog problems are now much cleaner.

During the last few decades, there has also been a growing consensus that we need to increase our awareness of how pollution affects our environment. Since 1997, vast resources have been devoted to studying environmental impacts, and we now have a host of data to consult when deriving conclusions. The problem becomes one of measuring environmental effects in an accurate manner. To understand the environmental impact, a few things need to be taken into account:

- **The magnitude of the problem:** Magnitude answers the question of how much impact is being caused. For example, how much pollution is being released via an energy process? How global is the extent of the pollution?

- **The severity of the problem:** Severity attempts to answer the question, how serious is the impact? This involves calculating, or estimating, the number of people affected, the extent of the damage, and how irreversible or recoverable the effect may be. Severity analysis also addresses impacts to the animal and plant kingdoms, although it's an unfortunate fact that humans take center stage in the drama. But that's beginning to change as more species become extinct, and those that are surviving are dwindling. People do care about animals and plants.

- **Whether the cause of the problem is a manmade or natural occurrence:** A variety of factors impact the environment, and while a problem may not be manmade, humankind can still play a role in mitigating the damage done. A case in point is the question about what causes global warming, which you can read more about in the next section. Say, for example, that scientists discover that sunspots are the cause. While there's nothing we can do about sunspots, we can take other action (like reducing greenhouse gas emissions) to avoid compounding an already existing problem.

Most impacts don't result in harmful effects, and many result in a better environment. Not all human activities are bad for the environment: consider forest management and river and stream damming.

Global Warming

The earth is surrounded by an invisible layer of insulation that regulates how the planet's temperature is maintained. The insulation is comprised of different gases — carbon dioxide (CO_2), methane, and nitrogen — that trap heat. These gases are all natural to the earth's atmosphere, and carbon dioxide in particular is a fundamental component.

Carbon dioxide is a natural element. While it is true that we humans release a lot of carbon dioxide into the atmosphere due to our fossil fuel combustion, it's not true that we're discharging a noxious manmade compound into the air.

While some may argue the primary cause of global warming (whether we humans or Mother Nature is to blame), there's little argument that global warming is actually occurring and that the rate of the warming is accelerating. The ten warmest years on record have occurred since 1990. In fact, global temperatures have not been this warm for over 1,000 years.

The central question about global warming is one of magnitude and severity (see the preceding section): How does the amount of carbon dioxide we release affect the severity of the global warming problem? The following sections delve into this issue.

Increasing carbon dioxide levels

We rely on greenhouse gases for our existence; they capture the sun's heat, thereby maintaining a livable environment. Without greenhouse gases, the average world temperature would be around 0 degrees Fahrenheit, which would probably preclude mankind's existence on the planet. What's going on now is that greenhouse gases are building up at an unprecedented rate, and we know for a fact that when greenhouse gases build up, the earth's insulation increases so that temperatures rise. There is no denying this cause and effect. Since the year 1000 ad, the average concentration of CO_2 in the atmosphere spiked sharply around the year 2000.

The most alarming aspect of the current situation is the acceleration in warming. In the late 1800's, worldwide temperatures were, on average, 1.3 degrees Fahrenheit lower than they are now. Some studies indicate that by the end of this century, temperatures will be, on average, three to nine degrees higher if the trend continues. Studies have also indicated the ocean levels have risen between four and eight inches over the last century. When the earth's temperature rises, ice packs melt and the ocean levels rise because there's simply more water. Also, when water temperatures rise, the volume of the water increases, so there are two effects from rising temperatures. These numbers are well above the averages of the previous century, but not unprecedented in the historical record.

In the last century, the atmospheric concentration of carbon dioxide has risen from 280 parts per million (ppm) to over 370 ppm now — the highest concentration in over 400,000 years and perhaps 20 million years. The International Panel for Climate Change (IPCC) projected that carbon dioxide levels will reach between 490 and 1,260 ppm by the end of this century. This is 75 percent to 350 percent higher than the concentrations measured prior to the Industrial Revolution.

Carbon dioxide lifetimes are very long, so this trend is deeply troubling because it implies that even if we curtail our carbon dioxide emissions, problems will remain for a long time.

And where is all this extra carbon emissions coming from? The Energy Information Administration (EIA) concluded that fossil fuel combustion is to blame for 81 percent of human induced carbon dioxide emissions in the United States alone. In some foreign countries, with lax pollution control standards and practices, this number is even higher. Table 7-1 lists the carbon dioxide emissions of commonly available fossil fuels.

Table 7-1	Carbon Pollution of Common Energy Sources
Energy Source	*Pounds CO_2/Unit*
Oil	22.4/gallon
Natural gas	12.1/British thermal unit
Liquid propane	12.7/gallon
Kerosene	21.5/gallon
Gasoline	19.6/gallon
Coal	4,166/ton
Electricity	1.75/kilowatt-hour
Wood	3,814/ton

Understanding the impacts of global warming

Global warming can affect both the human and animal populations in ways that can severely impact our way of life. While the exact severity of the impact is unknown, the following sections detail what you can expect — from dry spells to extinctions — if global warming continues.

Mounting violence in weather

The world has been witnessing increasingly violent weather events. Many people believe events such as the following are indicators of the precursors to the broader effects of global warming:

✔ **Heat waves:** In August 1993, Europe underwent an extreme heat wave, resulting in more than 35,000 deaths, particularly among the elderly, the young, and the infirm. In recent years, there are more deaths due to heat waves than the combined totals due to tornadoes, hurricanes, and floods. Heat induced strokes are going to become more common, as well as respiratory problems and heart disease.

✔ **Hurricanes:** In August 2005, Hurricane Katrina hit New Orleans and caused severe damage and loss of life. Meteorologists claim that severe weather events will become more frequent as a result of global warming. When huge masses of hot air collide with colder air descending from the north, the result is hurricanes and tornadoes. Meteorological records show that, in the last decade, there have been more severe hurricanes and tornadoes than the previous three decades combined.

✔ **Drought:** Around the world, droughts are increasing in frequency and severity. The United States, Africa, Spain, Chile, China, Australia, and other countries have experienced moderate to severe droughts within the last couple of years. While all droughts bring some degree of hardship, in less developed countries, they result in untold human suffering as water supplies dry up and food supplies disappear.

For the entire year of 2002, total damage from weather-related events worldwide exceeded $53 billion. In the United States, catastrophic weather related events have increased fivefold since 1970. There are no indications that these trends will decline, and in fact the evidence points to the opposite. Weather changes due to global warming are probably going to be as varied as they are widespread.

In addition, regional effects will be magnified: Drought regions can expect more severe and longer lasting dry spells. Wildfires will become more common, despite management efforts to control them. Flood-prone regions — where approximately 46 million people live — will be hit hardest because of the direct (the damage by the flooding) and indirect (the proliferation of diseases such as malaria and West Nile virus) effects.

Endangering wildlife

The animal kingdom will be hardest hit because they are much more exposed to environmental changes than are humans, who can control their environment. Mass extinctions will result from global warming, and biological breeding cycles will be altered, meaning reproduction will be altered.

Shifting agricultural centers

The silver-lining folks argue that global warming will alter crop growing and vegetation in such a way to increase production. But studies have indicated that a two-degree increase in temperature will leave plants drier and less viable, so the net aggregate production levels may actually decrease. At the very least, the agricultural centers will be geographically shifted, and this will result in massive relocations of farmers and agricultural workers to more northern climates that are now conducive to growing crops on large scales. This would reverse the southward bound population trend of the last few decades (which has, interestingly, been enabled by air-conditioning energy consumption). To be fair, it may be true that global warming results in more agricultural output because there will be more regions with weather conducive to growing different types of crops. But that helps humans, not the natural balance.

Melting ice shelves and glaciers

The recent phenomenon of melting icebergs and glaciers has been attributed to global warming. The U.S. Glacier National Park has seen a third of its glaciers disappear since 1850. Studies project that South America will see an 80 percent reduction of its glaciers in the next 15 years. Huge ice shelves broke completely free of Antarctica between 1995 and 2000. As ice melts, sea levels rise and coastal communities become displaced and the shorelines severely altered — an effect that tends to spread far inward as ecosystems inland from coastlines are very sensitive to the coastline conditions.

Rising sea levels

As the ice melts, sea levels rise due to the huge increase in oceanic water, plus the added effect of expanding water molecules due to the temperature increase. It only takes a small rise in sea levels before fresh water supplies become contaminated with salts and other sediments. This means water supplies for humans will dwindle, and in the face of increasing populations will result in severe political and social hardships.

Combating global warming

Most scientists who study global warming believe that it is humankind's greatest threat in history and is due to our use of fossil fuel. Despite all the dire warnings, and in the face of compelling evidence, many people believe that a wait and see attitude is warranted. The problem with this is that by the time the concrete evidence that they seek is available, it may be too late.

So the question is this: What do we do, as a society, about global warming? The most sensible answer is to continue studying and measuring the data, and to promote environmentally responsible policies.

Efforts to combat global warming have been initiated on a worldwide basis, with meager results. The international community drafted the Kyoto Protocol in 1997 and more than 170 countries signed on. Industrial nations agreed to cut their greenhouse gas emissions to 1990 levels, with a target date between 2008 and 2012. Developing countries were "encouraged" to cut their greenhouse gas emissions. Very few nations have met their goals. And in 2001, the U.S rejected the treaty, citing that its provisions put an inequitable burden on the U.S. economy.

Germany mandated a carbon tax on fossil fuels which accounts for around 11 cents per gallon of gasoline. German energy consumption is very low, per capita, and this is a direct result of their governmental regulations. Carbon sequestration, the process of capturing and storing carbon dioxide in old mines, is being pursued by the United States, Canada, Norway, and a host of other nations. Currently, however, the technologies are expensive and immature.

Ultimately, however, the real solution to global warming and carbon dioxide emissions is alternative energy.

Other Pollution Problems

Global warming gets all the buzz these days, but there are other environmental problems with fossil fuel emissions — ones that are much more direct and less subject to argument.

Smog

The problem of metropolitan air quality degradation has been well documented. The most common air pollution components are sulfur dioxide (SO_2), suspended particulate matter (SPM), lead (Pb), carbon monoxide (CO), nitrogen dioxide (NO_2), volatile organic compounds (VOC), ozone (O_3), and of course carbon dioxide (CO_2).

Smog is a reddish brown haze that settles over cities and valleys. It is the most visible manifestation of our pollution and in some regions serves as a visible, daily reminder that it's time to make some serious changes.

There is no predicting how smog may build up from day to day, or from month to month. Upper elevations where wind speeds are higher tend to have less smog, whereas valley bound cities get the worst of it because smog settles in and only grows worse. Different types of smog appear in different areas:

✔ **Sulfurous smog** appears in cold weather, like London and New York. Sulfur dioxides and particulates from electrical generation are the main causes.

✔ **Photochemical smog** is found in cities such as Los Angeles and Mexico City, and is far more common when it's warm and still. The main chemical constituent in photochemical smog is ozone (O_3), which is emitted by automobiles and trucks and electrical utilities as well as industrial manufacturing and in particular, oil refineries.

For the most part, smog is a local problem, but there is evidence that China's rampant smog is crossing the ocean to the United States, so smog is becoming a global phenomenon.

The pollution sources responsible for smog are stationary electric power plants and vehicular transportation. The two most important contaminants from motor vehicles that form smog are nitrogen oxides (NO_x) and volatile organic compounds (VOC). But volatile organic compounds come from many sources other than motor vehicles, many of these sources being natural parts of the environment. The formation of nitrogen oxides, on the other hand, results mainly from high-temperature combustion of fossil fuels. It is the nitrogen oxides that are the main controllable culprit in urban air pollution, and so these are being studied the most.

The density of smog in urban areas peaked in the year 1970, after which stringent air quality standards were implemented by the Federal Government. In particular, catalytic converters were required on all new cars. In California, even more stringent measures were adopted, and new cars sold in California include even better air quality technologies (California cars also cost more, and it is difficult to import cars from other parts of the country into California because they don't pass the state requirements). This demonstrates that governmental regulation can and does work, when implemented properly and with stringent oversight and enforcement.

Weather effects

Smog-laden skies affect the way heat is trapped in the atmosphere and as a result affect weather patterns. Rainfall shifts to different areas, and the natural ecosystem is altered. This affects wildlife and agriculture. Smog also weakens plants' immune systems, making them more susceptible to diseases and pests.

Health effects

The United States Environmental Protection Agency (EPA) lists the following health dangers from hazardous air pollutants (HAP) in smog:

✔ Neurological disorders

✔ Cardiovascular disorders such as heart attack and stroke

    ✔ Respiratory effects

    ✔ Liver disorders

    ✔ Kidney disorders

    ✔ Immune system deficiencies

    ✔ Reproductive deficiencies resulting in child development issues

    ✔ More than half of the HAPs cause cancer

Economic effects

Economically, smog and related air pollution issues cost a tremendous amount of money to a community. It is estimated that if Los Angeles would reduce its air pollution by 20 percent, it would save $9.8 billion per year in health and related costs.

Acid rain

Rainwater is critical to the earth's ecosystem. All living creatures rely on a consistent, clean supply of water. In the process of combusting fossil fuels, toxins are released into the atmosphere which react with water molecules to form acidic compounds. When these compounds come down in the form of rain, environmental havoc ensues. It's called acid rain.

In the natural state, rainwater reacts with carbon dioxide, but when human released levels of carbon dioxide increase, the process is accelerated and the natural pH balance of the ecosystem is altered. Unfortunately, a small alteration can cause big problems in the natural balance of our ecosystem.

Different areas of the country are affected in unique ways. The Midwest, for example, has a slightly alkaline soil that tends to neutralize acid rain, so this region is affected relatively lightly. In New England the effects are pronounced because the naturally acidic foundation tends to become aggravated by acid rain. Deciduous trees falter, pine trees whither, and overall agricultural productivity suffers. Certain mountainous regions in the Eastern U.S. have suffered the worst effects. Fish species are disappearing and those that remain are foundering. Certain varieties of trees, like the sugar maple, are disappearing because they can't handle the slight change in pH level.

Despite the local impact, global winds shift the acidic rainwater thousands of miles, so the problem is worldwide, not just local. In Canada, for example, a tremendous number of lakes are feeling the effects of acid rain.

Scientists have been monitoring the affects of acid rain since the 1970s, and the problem is getting worse each year. Acids have spread throughout the entire country, and even if the level of acid is reduced, the trend will continue because it takes Mother Nature decades to wipe the pH slate clean.

Acid rain can also affect humans directly. The acidic components are inhaled into the lungs and natural functions are altered in the process. Plus, acidic elements in soil and groundwater are ingested by humans and the resulting digestive problems can be pronounced. Declining bacterial populations in sewer systems are an issue because certain types of bacteria are necessary in the process of breaking down sewage into non-toxic elements. And finally, acid rain causes deterioration of buildings over time. Acid is very corrosive, and despite the fact that acid rain has very little acid, it's a persistent, consistent effect that accelerates the natural deterioration process.

All is not lost; improvements have been measured in acid rain levels, and the effects caused by acid rain. This goes to show that governmental inspired environmental programs can and do work, when implemented properly. Acid rain emission standards have improved the pH level of rain in many parts of the world. Yet there's a long way to go in the battle.

Increasing Demand and Dwindling Supplies

The need for alternative energy sources arises from two realities: increasing pollution (see the preceding two sections) and dwindling supplies of fossil fuels. Even if fossil fuel pollution levels could be brought under control (a dubious proposition in light of the developing world's reluctance to use pollution mitigation technologies), dwindling supplies will bring about a whole host of problems that could prove to be even more intractable than pollution effects.

In Chapter 4, I describe the ever increasing demand for fossil fuels worldwide, and in particular in the United States. Combine this with the fact that reserves of fossil fuels are dwindling, and you can see that we're simply going to run out of supplies, which will result in political havoc the world over, as countries scramble to meet their basic needs.

Alternative energy may not be able to completely offset the demand for fossil fuel energy sources, but it can put a very large dent in it. And as the use of alternatives grows, new technologies will come on line, displacing old technologies. As the trend grows, it will feed on itself and the growth will only accelerate. It remains to kick the alternative machinery into gear.

Political Movements and Political Ramifications

Inevitably, the solution to our fossil fuel problem will be driven by politics. Politics is a big part of the fossil fuel problem because government subsidizes the consumption of inexpensive fossil fuels. To counteract this promotion of carbon-based energy sources, political movements have arisen with the purpose of changing environmental policy, both statewide and national. Ultimately, it will be new technologies that change the face of our energy consumption patterns, but these new technologies are going to rely on political support structures.

The greens

Concern for the environment has grown into a formidable political force. The rise of green politics originated in the 1960s and 1970s. During that period, the Green Party became a fundamental force in Western European politics. In the U.S. there is no official green party, but there is a powerful green movement and it's become very influential in national politics. The green agenda seeks to reform government through environmental activism.

The greens have championed the development of alternative energy technologies. Green economics strives to include the true cost of energy sources in the dollar price that is paid for those energy sources. The need to include environmental costs is stressed, and this results in higher fossil fuel costs which tends to decrease demand. Hidden costs should be uncovered, and borne by those who use fossil fuels. Zero and low emission alternatives, in this paradigm, are actually less expensive than fossil fuels, and these will promote a cleaner environment that benefits mankind.

The New Environmental Paradigm (NEP) was crafted by the green movement to express their fundamental viewpoints:

- Human beings are but one species among the many that are independently involved in the biotic communities that shape our social life.

- Intricate linkages of cause and effect and feedback in the web of nature produce many unintended consequences from purposive human action.

- The world is finite, so there are potent physical and biological limits constraining economic growth, social progress, and other societal phenomenon.

The basic idea is this: Movement toward a new view of environmental justice should not be based on the consumer market. Instead, the real and direct cost of green energy should reflect the true cost for both the raw sources of energy, and the cost to the environment. When this is achieved, green energy will actually cost less than fossil fuels, and so the formidable power of the market will develop green alternatives. So the point is not to penalize fossil fuels unfairly, but to develop a fair and balanced accounting system.

Practical political advantages of alternative energy

Despite all the benefits that fossil fuel use has made possible, relying so heavily on carbon-based energy sources has put the U.S. in a rather precarious position today. We have to rely on not-always-friendly foreign governments for our fuel supply. We are vulnerable to price fluctuations due to geopolitical instability, environmental catastrophes, and other things beyond our control. And we are altering the environment in profound and often negative ways. All these things threaten our national security and our economic stability.

Pursuing alternative energy sources, therefore, has some very practical advantages:

- **Alternatives free the U.S. from interference by foreign powers that hold the world's largest fossil fuel reserves.** There is no telling how much it costs the U.S. both in terms of dollars and moral credibility when we order our political and military structures around the basic need of ensuring the consistent supply of foreign oil. Mideast politics has been incredibly volatile for the last several generations, and oil plays a major role in the equation. By using alternatives, renewables, and sustainables, the U.S. can wean itself from the foreign oil trap.

- **Alternatives are generally "local" energy sources.** When we rely too heavily on huge centralized energy generators (such as coal) our society becomes vulnerable to terrorism, catastrophic weather events, and economic perturbations like raw fuel supply disruptions. The only way a solar PV generator can be disrupted is if sunshine disappears. Same with wind and hydropower, not to mention biomass.

- **Alternatives are, in general, far more environmentally friendly than the fossil fuel mainstays.** The health costs of the various fossil fuel alterations of the natural environment are well documented. Health issues are very real, and costly in both dollar terms and quality of life.

- **The price of alternatives much better reflects the true, or real price, of energy.** Our current energy policies hide the true cost of fossil fuels, and make them far cheaper than they really are. If one is a true believer in the power of the marketplace, alternative energy is the clear winner.

✔ **Alternative energy is available to one and all.** Huge power plants take a tremendous amount of capital and decision making. Alternatives, on the other hand, can be pursued by everybody, on a basic level. The easiest way to pursue alternative energy is to place solar landscaping lights, which you can find for less than $5 at most hardware stores, in your yard.

A Final Thought on Fossil Fuel Use

Humans are invincible — we have our big screen TVs, our comfortable homes, and cushy jobs. This is what we've been striving toward for generations, and we've succeeded. All on the back of inexpensive energy, and in particular, inexpensive fossil fuels.

But for too long we've allowed ourselves to be lulled into a false sense of security. We have experienced natural gas and oil shortages, economic roller coasters due to price fluctuations, political and even military conflicts resulting from oil shortages, power blackouts and shortages, and declining reserves of new fuel deposits. We've experienced groundwater contamination, deteriorating ecosystems, agricultural decline in both quantity and quality, acid rain, smog, strange weather patterns. It's clear that we're affecting the earth in ways that we never even imagined only a decade ago. But we don't heed the warning signs because market forces tend to even things out, and we're only too willing to forget a crisis once it's passed.

In 1992, over 1,600 scientists came to this conclusion: "Warning to Humanity; Human beings and the natural world are on a collision course . . . that may so alter the natural world that it will be unable to sustain life in the manner that we know."

It's hard to argue with this conclusion, although some people do.

The fundamental problem is that most of us are not tuned in to how energy is produced or used. We take energy for granted. Fill up the car at the corner station, pay the utility bill, forget the rest of it. We haven't really had to pay much for energy, and the real price has been consistently declining so each year we're less and less inclined to pay much attention.

The challenges we face are symbolic of an even deeper problem: We've become unplugged from our fundamental roots. We've ignored Mother Nature and abused her to the point where she wants a divorce. We've developed a worldview that places humanity front and center as the only species that really matters (the rest you can go see in a zoo). Even in writing this chapter, I spent most of my time explaining how humans are affected by the impacts of fossil fuel use.

What we're most concerned with is the effects that environmental hazards have on humans, how they affect our own health. Extinction is a remote concept, despite the fact that it's occurring at an alarming rate the world over, but we tell ourselves things like, "Well, the dinosaurs went extinct and that had nothing to do with us."

We've lost our way. We've forgotten that we're a part of a huge, beautiful system not of our own making. We've become enamored of the concept that we're the masters, and because we're so good at making things, we've decided that we're the inventors of life on the planet earth.

It's time we wised up.

We rely on the earth's natural cleansing systems. We rely on the earth for everything that makes us tick and tock. When we alter the earth's ecosystem, we are altering humanity in the process.

But all is not lost, and it's certainly not too late to change course. Above all else, we humans have shown ourselves to be exceptionally creative and imaginative when it comes to transcending problems thrown our way. Our inventions and our acquisitiveness combine with our natural love of life to produce magnificent wealth and experience. So we'll figure out how to solve this energy problem we're immersed in. We all make our own little contributions, and they add up and grow and become infectious and at some point society as a whole will end up altered for the better. My writing this book, and your buying it and reading it, attest to the fact that people do care.

Part III
Alternatives — Buildings

The 5th Wave   By Rich Tennant

"Hold off on that. I think we're going to get solar panels."

In this part . . .

In this part, you can find out about the most important alternative-energy schemes used in residential and business buildings, such as nuclear power, solar power, hydropower, wind power, biomass and last, but not least, fuel cells — the shining promise of the future — in some people's eyes.

Chapter 8

Going Nuclear

....

....

The sun, which is the source of all energy on the planet earth, is a huge nuclear reactor. Natural nuclear reactors existed right here on the earth some 2 billion years ago and lasted for over 500,000 years before dying out. The entire universe is filled with stars, and each is a nuclear reactor. If all the energy on the planet earth comes from the sun, then all the energy on the planet earth comes from nuclear reactors. Sounds simple, but the story gets a lot more complex when we try to build manmade nuclear reactors.

Nuclear power is very clean, at least in terms of carbon dioxide emission, which is the cause of global warming. The fuel is relatively inexpensive as well, in terms of the amount of useable, or ordered, energy that it can yield per given weight. Yet nuclear remains controversial, although the underlying technologies are sound and well proven. It is undergoing a resurgence, however. In the coming decades, the number of nuclear power plants in operation is expected to increase by 25 percent or more.

In this chapter I describe nuclear technology, in layman's terms, and I describe the problems with nuclear power. While the technology is sound, there is still a lot of well-founded resistance. Many books do not include nuclear power as an alternative energy source, but I choose to because I consider alternative energy anything other than fossil fuel. This definition is predicated on the fact that it's fossil fuels that are causing the world so much harm. In relation, nuclear's problems are solvable, and manageable.

Getting to Where We Are Now

Upon its inception, nuclear power was advertised as the energy source "too cheap to meter." The original premise was that nuclear energy would be so inexpensive to create and dispense that it wouldn't even be worth the utility company's time to charge for it. In this scenario, the utilities would be run by the government, at taxpayers' expense, to provide all the electrical power needs of the country, indefinitely. Wouldn't that be great?

For two decades, nuclear mania spread across the world, but then it stalled due to engineering problems and system failures that resulted in panics and political confrontations, sometimes violent.

Nuclear power: The early years

During World War II, the United States launched the Manhattan project, which produced the first atomic bombs. With technological advancements based on this fundamental research, in December 1942 the first nuclear reactor was tested in an abandoned handball court at the University of Chicago. This test proved the viability of large scale nuclear power reactors for utility scale use, and the race was on.

In 1946, Congress created the Atomic Energy Commission (AEC) and chartered it to regulate the development of large scale nuclear power for use on a commercial basis. In 1951 an experimental reactor in Idaho produced enough power to prove that nuclear technology is sound enough for practical and widespread applications.

Protests and accidents in the '60s and '70s

Throughout the 1960s and early 1970s there was a lot of contention over the safety of nuclear power plants for commercial use. Court cases were filed and delays were experienced, and some reactors that had already been started were delayed. In 1971, for example, the Calvert Cliffs Coordinating Committee (CCCC) won a lawsuit against the Atomic Energy Agency (AEA) and a reactor scheduled to be built in Maryland was delayed for a number of years. Construction of subsequent plants around the world were delayed or cancelled, as a result.

In Germany, an international movement to halt atomic energy crystallized, and the movement spilled over into many other countries, particularly in Europe. Some of the protests erupted in open violence. At Brokdorf in 1981, 100,000 demonstrators confronted 10,000 police at a nuclear building construction site. The demonstrators were armed with gasoline bombs, sticks, stones, and slingshots, and 21 policemen were injured. There is no denying that the word "nuclear" elicits strong sentiments, rightly or wrongly.

Plant orders peaked in 1974 then fell off sharply after 1974 when energy-conservation sentiments grew as a result of the Arab oil embargo. The feeling was that the world could be saved by conservation and efficiency alone, although that hasn't proven true in any sense of the word (as I explain in Chapter 3). The anti-nuclear forces sought not just to stop nuclear power, but to completely eliminate it as an option.

Political and environmentalist support for nuclear took two huge hits due to a couple of widely publicized nuclear reactor accidents:

- **Three Mile Island:** In 1979, at Three Mile Island in Pennsylvania, an enclosed reactor failed when internal temperatures exceeded 2,750 degrees Celsius. This could have resulted in a terrible explosion, but did not. The containment building, which was built specifically for such mishaps, prevented any immediate deaths, and contained the radiation leakage to minimal proportions. Studies indicated, however, that the cancer rate increased slightly amongst the nearby population and public fear was stoked.

- **Chernobyl:** In 1986, in Chernobyl, Russia, a nuclear reactor with substandard engineering and a poorly designed containment structure melted down and released radioactivity. The accident exposed thousands of people to high levels of radiation, particularly during the cleanup (we have a special job for you, comrade!). A 300-square-mile area around the site was evacuated, and there were 31 immediate deaths, with 500 more people hospitalized for various reasons. It is estimated that between 6,000 and 24,000 people died from cancer since the accident.

Since these accidents, support for nuclear energy has waned considerably. The last nuclear plant built in the U.S., for example, was finished in 1979, the same year as the Three Mile Island nuclear power plant disaster. The valve was shut off completely, for better or worse.

The situation today

In November of 2007, NRG Energy, in Princeton, New Jersey, became the first company to file for a license to build a new nuclear plant since the 1970s. A dozen more applications have subsequently been filed, and more are expected.

This resurgence is a direct result of global warming fears, and the fact that nuclear power plants do not discharge greenhouse gases. As I explain in Chapter 18, with the development of the electric car, worldwide demand for electricity from the grid is expected to surge, and nuclear provides the cleanest option (simply displacing internal combustion engines with electric motors does not solve the greenhouse gas crisis, it merely moves the burden onto electrical power plants, most of which are coal-fired).

At present, nuclear power plants provide around 21 percent of U.S. electrical energy. France, which has promoted and subsidized nuclear for decades, derives over 75 percent of its energy from nuclear, and there have been no major accidents or environmental mishaps. Japan gets 24 percent of its electrical energy from nuclear, while Germany made the decision in 2000 to phase out of nuclear entirely (this may have more to do with the politics in Germany, post World War II, than any pragmatic analysis of nuclear power's role in a diversified energy economy). Table 8-1 shows the nuclear power current capacities of different countries.

Table 8-1	Nuclear Power Capacity, by Country		
Country	*Capacity (GW)*	*Share of electricity (%)*	*# reactors*
U.S.	98	21	104
France	64	78	59
Japan	45	25	54
Russia	21	17	30
Germany	21	28	18
South Korea	16	40	19
Canada	12	12.5	17
UK	12	24	27
China	12	unknown	15

The Fundamentals of Nuclear Power

The possibility of nuclear energy of the type we use today in power plants was illuminated by Einstein's famous equation:

$$E = mc^2$$

where E is energy, m is mass, and c is the speed of light (around 3×10^8 meters per second).

To put this into perspective, this equation says that an apple has enough mass to run New York City's electrical needs for an entire month. So there is considerable energy potential in nuclear power, and the amount of raw fuel consumed is minimal by fossil fuel standards. It does sound great in theory, and it's easy to see why early enthusiasts championed nuclear power to such an extent (refer to the earlier section, "Getting to Where We Are Now").

There are two distinct types of nuclear energy: fusion and fission. When atomic nuclei undergo either fusion or fission, it is the binding energies that provide the useable power output. In fusion, atoms are combined, or fused. In fission, atoms are dissociated. In both cases, the basic physics exploits Einstein's famous equation. It may seem like magic, but when nuclear reactions occur, mass combines to equal less mass, and the difference is converted into radiation energy.

The simplistic physics maxim that mass is conserved is not quite correct. But the first law of thermodynamics is correct: Energy can neither be created nor destroyed. According to Einstein's equation, mass is just another form of energy. So in a nuclear reaction, it *is* energy that is conserved.

Atoms

All matter is made up of very dense particles called electrons, protons, and neutrons (these, in turn, are made up of even more fundamental particles called quarks and gluons and all sorts of other strange things). The subatomic particles move incredibly fast and electrons are in motion most of the time around the nucleus, which is made up of the protons and neutrons, as Figure 8-1 shows.

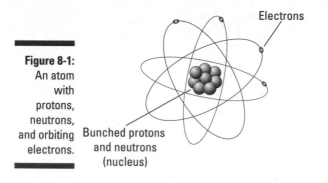

Electrons

Bunched protons
and neutrons
(nucleus)

Figure 8-1:
An atom
with
protons,
neutrons,
and orbiting
electrons.

The number of protons and neutrons defines the properties of the element, and is called the *nucleus*. Protons have a positive charge, neutrons have no charge, and electrons have negative charge. In most atoms, the number of protons, neutrons, and electrons are equal, and the net charge is zero because the positive charges balance out the negative charges; however, it's not uncommon for an atom to contain different numbers of particles.

The number of protons determines the *atomic number* of the atom, and this gives the element its fundamental identity. The simplest element is hydrogen, with one proton in the nucleus. Helium has two protons and lithium has three. The periodic table of the elements (remember from Chemistry 101?) lists all the known elements, with increasing atomic number. Elements with high atomic numbers are more dense than the simpler elements.

Nuclear binding energy is the "glue" that holds protons and neutrons together, forming a nucleus. Nuclear binding energy implies that a force is holding nuclei together, for most nuclei consist of a number of protons, each with a positive charge, and like charges repel. In other words, two protons repel each other with a considerable vigor, so the fact that protons are found bound together suggests that a force much stronger than electrical repulsion "binds" nuclei. The force must:

✔ Be strong enough to overcome the repulsion forces of the charged particles.

✔ Be extremely short in range because once nuclei particles are separated, they do not rebind spontaneously.

Isotopes

Elements can have different versions characterized by slight changes in the composition of the nucleus; the different versions are called *isotopes,* and despite the fact that the differences are minor, the way isotopes react with other elements can be major. Isotopes occur when the number of protons does not match the number of neutrons. All elements have at least two isotopes, and some have dozens. Each of the isotopes will react differently, and scientists can control the composition of atoms in order to exploit the different reaction potentials.

Elements all have a particular isotope that's found in its natural state on the planet earth. Some isotopes are very stable and retain their form over long periods of time. Others tend to decay into more basic, stable forms, or they decay into completely different atoms. When nuclei decay spontaneously (some nuclei are right on the brink of changing into a different form and a tiny tweak of energy will make them tumble over the edge while other nuclei are relatively distant from changing into other forms), they are referred to as unstable. The decay is always accompanied by some kind of energy release, or absorption. When an unstable nucleus decays, the energy is called radio-activity, or radiation, and this is of much concern in nuclear power because radioactivity can be very hazardous (I get into this in more detail in a subsequent section).

Uranium is the element used in nuclear reactors, and all known isotopes are unstable. The rate of decay varies from one element to another, and some elements can take thousands of years to decay, but decay they will.

Atomic weight

By definition, the atomic weight of an element is equal to the sum of the number of protons and neutrons in the nucleus. Carbon, which is a very common atom in the earth's ecosystem, has an atomic weight of 12, meaning the nucleus is composed of six protons and six neutrons. This particular version of carbon is called carbon-12, or C-12. Some carbon atoms, however, contain eight neutrons in the nucleus instead of six, and the total protons plus neutrons equals fourteen. These atoms are called C-14.

Uranium exists as either U-234 or U-235 (with atomic weights of 234 and 235, respectively). The most common form of uranium, however, is U-238. Nuclear power plants use U-235 almost exclusively because it's easy to induce controlled decay and thereby create a consistent, steady stream of output energy in the form of heat radiation.

Ions

If an atom has more or fewer electrons than protons, it's called an *ion*, and there is a net electrical charge. When there are more electrons than protons, the net charge is negative; more protons than electrons results in a positive net charge. It's still the same fundamental element, but its properties become much different. In particular, an ion is capable of electrical interactions that may be exploited to create electrical flow. When a substance, whether solid, liquid, or gas, contains mostly ions, it is said to be *ionized*, and electrical flow is a natural consequence of the physics.

Elements can be isotopes and ions at the same time, and this creates even more opportunity to induce energy flow and creation.

Nuclear Power Today: Fission

I spent eight years studying physics, and I will have to admit that nuclear physics is a tough nut to crack. But don't worry; you can understand some basic premises, and in the process you'll know what makes nuclear reactors tick and why they tick so loudly. Forge on, and it will be worth the effort because nuclear power, while not magic, is awe inspiring. There's an inherent beauty in the atomic world. In the following sections I describe how nuclear reactors work and some of the problems that are inescapable. Getting reactors to work is not what's causing so much controversy; it's the solution to the problems.

How nuclear reactors work

At the core of a nuclear reactor are U-235 atoms, and these are bombarded by high speed neutrons. When an atom of U-235 is struck by a neutron, the following happens:

1. **The atom instantly splits into two separate nuclei, each lighter than the original.**

2. **Neutrons are emitted, along with gamma rays, which are extremely energetic photons that can penetrate most substances without hindrance, and heat energy is produced.**

   Heat energy is what we are seeking in a controlled nuclear reaction in an electrical power plant. As heat energy is released, the uranium itself heats up.

3. **When one of the emitted neutrons from the original nuclear reaction strikes another uranium atom, the new uranium nucleus splits, thereby releasing even more energy, and creating new atoms as a byproduct.**

4. **A chain reaction occurs when enough neutrons are being emitted by the fission to continue inducing new nucleus decays on a steady basis.**

As the chain reaction continues, uranium gets used up and the chain reaction begins to die out. At some point, there are not enough new neutrons being emitted to keep the reaction going. So there is a need to constantly keep feeding in new uranium material in order to create the consistent, steady flow of energy that is desired from a nuclear reactor. There is also a need to remove the decayed byproducts, and this can be a real problem. The decayed byproducts are called *spent nuclear fuel.*

One of three states are feasible when a mass of uranium is subjected to controlled neutron bombardment:

- ✔ **Subcritical state** occurs when the reaction dies down because there are not enough new neutrons being created as uranium atoms undergo fission to continue inducing new reactions. The process simply dies out and eventually stops.

- ✔ **Critical state** occurs when the reaction is self sustaining, meaning that just enough new neutrons are not only being created, but finding new uranium atoms to strike and induce another fission event. This is the goal of nuclear reactors, and a delicate balance is achieved in the core of the reactor that maintains the critical state.

- ✔ **Supercritical state** occurs when the reaction increases to the point where so much heat energy is being created that the uranium core melts. This is very dangerous, and is the classic nightmare of nuclear energy opponents. It's referred to as "meltdown," and is the subject of popular movies and science fiction novels.

A single pound of enriched uranium (not an easy pound to get, considering all the enrichment needed) yields around 33 billion Btus of energy. However, the process is not particularly efficient, so to give an idea of the practical output of nuclear power, one pound of uranium is equivalent to around 3,300 gallons of gasoline, 23 tons of coal, or around 134,000 kWh of electrical power. That's a lot of energy packed into a small package.

Radioactive decay

Radioactive decay is a spontaneous disintegration of a nucleus which releases different forms of energy. The process is described by

A decays into B and b

where A is the radioactive nucleus, B is the resulting nuclide, and b is emitted radiation.

The rate of decay of a substance is important because it describes the relative safety of the substance. When radioactive decays are fast, the process is over quickly. When radioactive decay takes a long time to exhaust, the material is unsafe for thousands of years.

There are four types of radiation in the nuclear processes that we use for commercial applications: alpha, beta, gamma, and neutron.

- ✔ **Alpha:** Low level alpha particles travel only a few inches in air before dying out. They are incapable of penetrating a sheet of paper, or human skin, so the risk is minimal.

- ✔ **Beta:** Beta particles, or electrons, are significantly more penetrating, and can affect the outer layers of human skin, as well as internal organs. However, they can only travel a few feet before combining with natural elements at which point they become harmless.

- ✔ **Gamma:** Gamma rays are highly destructive to biological systems. They travel at the speed of light, and are highly penetrating since they have such high energy levels. Dense lead is used to prevent the travel of gamma rays (which are similar in nature to X-rays).

- ✔ **Neutron:** Neutron radiation poses very little threat to living organisms because it occurs at the core of nuclear reactors, and quickly dies out due to absorption by the nuclear process. Containment is easy enough.

All living creatures are constantly exposed to radiation in low levels. It's a fact of our natural environment. Many of the earth's materials release forms of radiation, and in fact the earth's ecosystem relies on certain radiation processes to function normally. Without naturally occurring radiation, the world would be a far different place.

Americans are naturally exposed to over 200 mrem (millirems) of radiation per year. In some parts of the world that number is much higher. In Ramsar, Iran, natural radium in the ground exposes nearby residents to over 26,000 mrem per year, and there is no record of increased cancer rates there. Radiation levels near nuclear sites never exceed the 400 mrem. This is a real problem for nuclear power opponents who claim that radiation will cause increased disease and sickness near power plants. The record simply doesn't bear this out, and since natural levels are so high in some areas (without dire consequences), the paranoia is unjustified.

Radiation is also emitted by microwave ovens and X-ray machines, and people are hardly afraid of these manmade sources (well, some people are).

All forms of radiation kill cells in a living organism, not just the types that are emitted from nuclear reactors. But biological entities continually fight off radiation sources, most of the time very successfully. In fact, our immune systems are strengthened by the low level battles fought against naturally occurring radiation. It's like exercise — it creates strong muscles.

However, when doses of radiation become inordinately large, there are a number of system failures that can result in death and certain types of cancer. Too much goes way too far. The greatest health problems occur when radiation is ingested via food, water, or air. When too many cells are damaged in too short a time, individual organs may shut down entirely, or the complete system may simply decide to give up the ghost.

Nuclear waste

Nuclear waste is a problem, there's no denying that. But it can be safely managed, and it has been. In American reactors, the nuclear fuel rods need to be replaced every 18 months. When they are removed, they contain large amounts of radioactive fission products and they're still hot enough that they need to be cooled in water. As time progresses, the radioactivity levels decrease, and the heat also decreases. It's the nature of all nuclear waste that as time goes by, it becomes less dangerous. The question is what to do with the waste as it grows progressively weaker.

This has been a subject studied intensely for over 50 years now. The amount of nuclear fuel consumed in a reactor is termed *burnup* and is expressed in units of megawatts-days per metric tons (MWd/Mt). For an average 1000 MWd reactor containing 100 Mt of fuel, operating at around 33 percent conversion efficiency of thermal to electric, burnup is around 33,000 MWd/Mt. The burnup would consume 3.3 Mt (metric tons) of uranium fuel, or around 3 percent of the fuel in the reactor core. The spent fuel would contain 97 Mt of processed nuclear fuel, along with 12 kg of fissionable plutonium, and 3.3 Mt of high energy radioactive waste. This is an obvious problem.

There are basically three options with regard to disposing of the spent fuel:

- **Keep it in special storage chambers at the nuclear reactor sites.** This creates local hazards and makes security a prime concern at the sites. In the United States, this is standard procedure because environmentalists have stymied all efforts to transport the material to a central waste site (their concern is that in transit, an accident may occur which spills nuclear waste).

- **Store it in "temporary" disposal sites, in New Mexico and Nevada (such as Yucca Mountain).** This makes the most technical sense, but there is considerable political resistance so plans have been postponed. Ultimately, this is going to happen, however, since there simply isn't a better solution.

- **Reprocess it to extract new, recycled fuels.** More than 95 percent of the material in a spent fuel rod can be recycled for energy and medical isotopes. In much of the world, this is done routinely, but in the U.S. this solution was halted for political reasons.

Stopping proliferation?

In the U.S., reprocessing was halted during the 1980s based on the unfounded belief that other countries would stop as well, and the supply of nuclear materials for use in weapons would be halted as a consequence. Because of this chimerical belief, the U.S. now has more nuclear waste than any other country in the world, at around 144 million pounds. It should be mentioned that a lot of this waste has resulted from weapons research and production, not just power plant waste.

As for nuclear weapon proliferation: Every country that has sought nuclear weapons has succeeded in obtaining them. The existence of nuclear power plants in western countries has had no effect, one way or another, on this trend. The way nuclear fuel is used in western reactors in fact makes the use of waste products for weapons impossible. So the U.S. decision to halt reprocessing, in hindsight, was disastrous in light of the consequences.

In most of the rest of the world (other than the United States) reprocessing extracts useable uranium and plutonium from the waste materials. As a consequence, the remaining materials are very weak, and storage is a relatively easy matter. By 2040, for example, England will have just 70,000 cubic feet of this weakened waste material. This is the volume of a cube with sides only 42 feet. The British government has determined that "geological disposal" (burial deep beneath the ground) is the best option.

France has proven that reprocessing works. With a fully developed nuclear power capacity, they lead the world in providing clean energy to their citizens. The French now store all the waste from 30 years of producing 75 percent of their electricity beneath the floor of one room at La Hague, in Normandy.

After ten years, nuclear waste products are only one thousandth as radioactive as they were initially. After 500 years, they are less radioactive than the uranium ore they originally came from. The waste question is simply one of storage; the technology is sound. The dispute is all about politics.

Uranium mining

Producing the U.S.'s nuclear energy takes around 52 million pounds of uranium, which is mostly imported from foreign sources (77 percent of the total). The primary foreign sources are Canada, Russia, Australia, and Uzbekistan. However, obtaining nuclear raw materials from foreign sources does not entail the same types of problems inherent to foreign fossil fuel sources since there is much less raw nuclear fuel being used, and the source countries are of a much different political complexion than the sources for fossil fuels.

Uranium, a heavy ore mined from the ground, can be found over most of the globe in trace amounts, and this is a source for some of the natural background radiation of the earth. Most deposits are too low in density to extract economically. Over 99 percent of the naturally occurring radiation is U-238, which is unsuitable for use in nuclear power reactors. Less than one percent of the world's uranium comes in the form U-235, which is the isotope used in nuclear fission power plants.

There are three methods of mining:

- ✔ Over half of all uranium is mined from large underground caverns, similar in nature to coal mines (although not as dark and dreary).

- ✔ Another 27 percent is obtained via a range of surface mining techniques, which leaves bad scars on the environment, not to mention high radiation levels in the surrounding area.

- ✔ In-situ leach mining accounts for the remaining 19 percent. This process uses oxygenated water to dissolve the uranium without disturbing the ore itself. The water is pumped up to the surface and processed to extract the uranium. Present day mining operations target ores that are at least 0.1 percent uranium, and there is a lot of waste material consequent to the process.

A substance called yellow cake is produced from the raw uranium. In-situ mining produces yellow cake on site, without the need for special refining processes. The yellow cake is then dried and formed into fuel rods, which are the raw fuel used in fission power reactors. Needless to say, this material is very hazardous, and expensive. Never order yellow cake for dessert.

The design of today's reactors

With a very small amount of uranium, nuclear reactors of the fission variety (the only kind that are in existence today) can create huge amounts of useable heat energy, for a long, long time. In general the process works like this:

1. **A core of uranium is maintained in the critical state (the self-perpetuating chain reaction described earlier).**

2. **Heat from the nuclear reaction is channeled into the boiler (this is similar in nature to the boilers used for fossil fuel combustion power plants).**

3. **This heat is converted into steam, and the steam powers a huge turbine/ generator assembly which provides electrical power to the utility grid.**

4. **A condenser converts the used steam energy back down to a liquid, which is pumped back through the boiler for further production of electrical energy.** Some reactors don't allow the water to boil; they keep it under high pressure and use that pressure to spin the turbine.

   Note: The boiler and heat transfer portion of the system are entirely unconnected to the nuclear reaction core so that the liquid used for boiling and condensing does not become radioactive.

   In a boiling water reactor, the heated water is allowed to turn into steam prior to powering the turbine/generator. In a pressurized water system, the water is contained so that it cannot boil — instead it gets super-heated and pressurized.

A number of different types of reactors have been built on the principles of the basic design described above. By far the most common type is the light-water-cooled reactor. Water is used to carry the heat from the reaction to the turbine/generator, and it's also used to cool the core. There are two types: boiling water and pressurized water, with the latter comprising two thirds of worldwide production. Other options have been pursued, and they work well enough, but there is something to be said for technological maturity, and since the light-water-cooled reactors are well known they are the favored choice.

Designed for safety

To ensure safe operation, safety precautions are designed to ameliorate the problems inherent in nuclear power via fission and take a variety of steps, including the following, to ensure the safe production of nuclear power:

✔ Housing the core within very thick, dense structures so that emissions of radiation and radioactivity are well contained. The radiation shield prevents the emission of radioactive byproducts into the surrounding environment (extremely important). The containment vessel also prevents the escape of radioactivity, as well as heat.

✔ Housing the entire system in a secondary containment structure, as an extra precautionary measure that is generally spherical in shape. Spherical shapes can better withstand external catastrophes such as a terrorist attack, tornadoes, earthquakes, and the like. American containment structures can withstand the impact of a jet fighter going full speed without a breach in the seal.

✔ Maintaining precise temperatures in the core by carefully pumping huge amounts of cooling water throughout the system.

✔ Ensuring that materials used in the reaction meet predetermined purity and density specifications.

There is actually very little chance of a nuclear explosion in a nuclear power plant. There are plenty of problems that can occur with a nuclear reactor, but a nuclear explosion is not one of them. If a reactor does go "supercritical," the uranium at the core melts due to the excessive heat. But reactor-grade U-235 is not processed enough to initiate the violent chain reaction at the core of a nuclear bomb. However, meltdowns can contaminate groundwater and air with radioactive emissions.

Advantages of fission reactors

There are a number of advantages of nuclear fission electrical power:

- ✔ Uranium is relatively inexpensive, in relation to fossil fuels. The energy density is extremely high so the amount of raw fuel is much less than for fossil fuel counterparts.

- ✔ Uranium is found in many different parts of the world, so supply issues are minor in relation to fossil fuels.

- ✔ If push came to shove, reprocessing the current waste stockpile could provide raw fuel.

- ✔ Maintenance of a nuclear power plant is infrequent, compared to fossil fuel type plants. Once a nuclear power plant is up and running, the chain reaction can be maintained for long periods without outside interference.

- ✔ Fission reactors do not need oxygen to operate, which means they can be sealed very thoroughly from the outside ambient. This prevents air contamination from the radiation that is released in the reaction.

- ✔ Nuclear reactors can be built underground, which enhances their safety because the ground acts as a huge containment vessel.

- ✔ Fission power reactors do not emit carbon dioxide, or other pollutants associated with fossil fuel combustion.

Disadvantages of fission reactors

Of course there are a number of disadvantages as well:

- ✔ Nuclear reactors cannot be easily turned up and down. They want to output consistent amounts of energy and are usually turned up to around two thirds capacity or more. When load requirements vary, backup power sources need to be employed to deal with the changing loads. This is fine, most of the time, because utilities have a very good idea of how much power is going to be needed at any given time.

- ✔ Uranium mining is dangerous because it exposes workers to potentially lethal doses of radioactivity. It also releases radioactivity into the air, and this can be swept great distances by the winds.

 ✔ Fission reactors produce a lot of nasty waste products, and these are dangerous.

 ✔ Although small, there is a risk of terrorist attacks because of the high containment. If a terrorist gets a hold of the nuclear waste products, they could be used to sabotage an entire community by "dirtying" the water and air supplies.

 ✔ Moving fissionable materials is hazardous, both in supplying the reactor with fresh uranium and in carrying away the disposal products. Accidents do happen, and an accident involving nuclear products could have widespread ramifications.

Hydrogen Fusion: Nuclear Power of the Future

The sun provides its energy via hydrogen fusion, which requires extremely high temperatures. In the sun's case, this high temperature occurs because of the tremendous gravitation pull at the core due to the mass of the sun. The compression maintains a hot enough temperature for a constant nuclear fusion reaction to occur.

Fusion promises great advantages over fission because the only byproducts are naturally occurring elements on the earth. There is no radioactive danger from controlled fusion reactions.

The process at a glance

Hydrogen fusion occurs in several distinct steps (shown in Figure 8-2):

1. **Two hydrogen nuclei (protons) combine and emit a positron, which has the same mass as an electron, but positive charge, and a *neutrino*.**

    Neutrinos are similar to electrons and positrons but have tremendous energies that enable them to penetrate very dense materials.

2. **One of the original protons changes state into a neutron.**

    The resulting product is a deuterium atom (H-2), which is a heavy isotope of hydrogen composed of one neutron and one proton.

3. **The deuterium atom combines with another proton to create helium-3 (H-3), which contains two protons and one neutron, and emits a burst of gamma radiation.**

   4. **Two separate H-3 nuclei fuse together to form He-4, which has two
      protons and two neutrons. In this phase, two protons are rejected, and
      these feed further reactions that start this entire process over again.**

In the end, the total mass of the matter produced is slightly less than the
total mass of all the ingredients, and this is the source of energy.

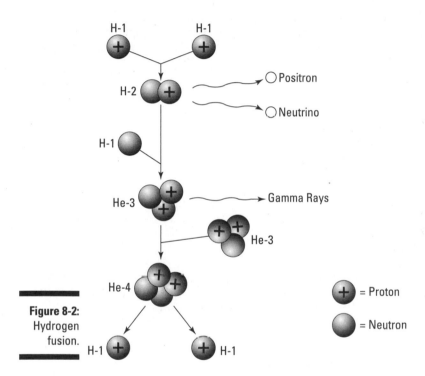

Figure 8-2:
Hydrogen
fusion.

There is enough hydrogen in the sun to keep its atomic chain reaction going
for millions of years, and the sun has already been burning steadily for millions of years. When distant stars die out, it's because the supply of hydrogen
is depleted.

There are currently no working fusion power reactors, but there are enough
advantages of using this type of nuclear reaction over fission that a lot of
research and development money is being spent on finding solutions.

Atomic bombs

In a hydrogen bomb, a slightly different reaction is induced. The chain reaction is nearly impossible to control, but that's not an issue with a bomb (in fact, that's the desired end result). Here's the reaction at the core of an atomic bomb.

Tritium (H-3 — one proton and two neutrons) combines with a nuclei of deuterium (H-2 — one proton and one neutron) to form He-4, or helium, along with an ejected neutron. A great deal of energy is also emitted in the process, since the mass after the combination is less than the mass before.

In a working bomb, the heat that induces this reaction is supplied by a small fission reaction (which is easier to induce and control than fusion). The fission reaction burns out very quickly, but while it lasts there's enough energy to induce a tremendous fusion reaction. A fusion reaction burns out when all of the core material is exhausted. But in the process, there's an atomic explosion, and we all know what that means — mushrooms!

Advantages and disadvantages of fusion

A list of the advantages of nuclear fusion illustrates why it could be a large part of the solution to our current energy quandary. In fact, many people believe that fusion power, combined with fuel cells, is the ultimate solution to humankind's energy problems.

Here are the advantages of nuclear fusion power plants:

✔ The main byproduct of a hydrogen fusion reaction is He-4, which is a harmless gas. Tritium is also emitted, but that can be used as a clean fuel (I won't get into the details here, which are very technical). As opposed to fission, fusion is much cleaner on the environment.

✔ Deuterium, the original fuel in the fusion reaction, can be easily manufactured from water, which is obviously in abundant supply.

✔ Deuterium and tritium are the only fuels necessary for a fusion reactor, and both are relatively cheap and abundant.

✔ There are no greenhouse gas emissions from a fusion power plant.

✔ A working reactor will actually be safer to operate than a fission reactor because meltdown cannot occur if the reactor core is damaged. Fusion reactions cannot be sustained without a continuous flow of new fuel, and this is straightforward to control.

✔ Fusion is not a chain reaction, so it cannot self sustain. The reason a bomb works is because there is enough fuel to ignite the explosion, but then the fuel is exhausted very quickly. In a fusion reactor the amount of fuel can be controlled very carefully so that there is not enough at any given time to cause a catastrophic event.

As with everything under the sun (back to our original fusion reactor, the sun), there are disadvantages of fusion power:

✔ Despite the fact that there are no radioactive wastes, as there are with fission reactors, the emitted neutrons eventually cause too much radioactivity in the core for the process to be safe any longer. This problem can be solved, but at a very high cost.

✔ The widespread deployment of fusion reactors is not going to happen for some time.

✔ The public is very wary of any technology that operates basically as a controlled nuclear bomb. The political hurdles are formidable, and as fission has shown, political hurdles are numerous and powerful.

Issues Impacting the Move toward Nuclear Energy

With safety concerns front and center, the true question as to whether nuclear power should be pursued is one of economics and politics. Nuclear power technology has been well proven over the last 40 years, and there have been only two major nuclear power reactor problems (Three Mile Island and Chernobyl). The overwhelming vast majority of nuclear power reactors operating in the world have been extremely reliable and safe. The nuclear record is, in fact, much better than the fossil fuel record. Yet there is a stubborn mythology associated with nuclear power.

It's unfortunate that the same word used to describe atomic bombs and nuclear Armageddon has to be associated with nuclear power. It's like associating all fossil fuel consumption with Molotov cocktails. Imagery and symbolism are very powerful elements in human perception. In this section I address some of the more persistent elements of nuclear mythology. Some may accuse me of taking a political stand here, and perhaps I am. Nuclear is going to play a role in the world's energy drama more and more as time progresses. For that to happen, people need to understand what's myth and what's not.

The fear factor

Nuclear reactor safety has been a big issue in the entire 50 year history of the technology's widespread use. There are a lot of unfounded fears, none of which has proven valid. Unfortunately, these fears have prevented more widespread use of nuclear power, the consequences being that our greenhouse gas problems are worse than they need to be. Here is a short list of the main targets of the hysteria:

✔ Reactors can explode like an atomic bomb. There is no physics basis for this conclusion, and the engineering details of western power plants preclude anything like this from happening. The problem is that an exploding power plant is easy to visualize, while the physics that proves that it's impossible is very difficult to understand. Simple images win the day.

✔ A massive release of radioactive elements could occur due to an accident at a nuclear plant. It's true that in Chernobyl there was a tremendous radioactivity problem but the plant was built substandard, without the kind of containment technologies used in Western countries. Chernobyl only proved that foolish nuclear energy is very foolish indeed.

The Chernobyl accident was caused by failures of Soviet style reactor systems. When Soviet style reactors get too hot, the rate of the nuclear reaction increases, which is the exact opposite of what happens in western style reactors. Plus, the Soviet reactors don't have containment vessels to prevent radiation from escaping. And the area surrounding Chernobyl has now become a haven for wildlife, disproving the nuclear winter theory. The point is that little of value was learned from Chernobyl, aside from the already obvious fact that Soviet technology was way behind western technology.

✔ Meltdowns can occur due to a simple loss of coolant liquids. Coolant liquids maintain temperatures at the core of the reactor so that critical state can be maintained. When coolant liquids disappear, temperatures rise, but there is not enough raw reactor material to cause the reaction to go supercritical. Meltdowns occur because there is too much reactive material, and this is well controlled by safety features not related to coolant liquids.

✔ There is a continuous release of radioactivity under normal operating use. In other words, if you live next to a nuclear plant you may end up glowing in the dark, or having two-headed children. The fact is, there is more radioactivity released from a coal mine site than from a normally operating nuclear power reactor.

✔ Waste heating will be ecologically damaging. This is no more true than for fossil fuel combustion power plants. Take your choice; either way you get the heat. But waste heat is not nearly as environmentally damaging as pollution byproducts.

✔ An accident will occur during transit of radioactive materials. This has never occurred, and if an accident does occur, containment vessels are rugged enough to withstand the tremendous forces that would be expected in an accident.

✔ Radioactive materials may be used by terrorists. This is true, although there are a huge number of safeguards to prevent it. But terrorists have proven that they don't need nuclear materials to wreak havoc.

✔ Disposal of radioactive wastes can never be satisfactorily achieved. In fact, radioactive wastes have been contained successfully for the last 50 years, with no record of environmental exposures over the natural state of affairs. People tend to forget that radiation occurs naturally in every part of the world. Nuclear plant radiation emissions are low compared to the natural state in many parts of the world, and the dire warnings of rampant cancer have never come true.

While there may have been good reason for the cautionary tone of the 1970s because nuclear power was a very big question mark, past predictions of catastrophic accidents haven't even come close to the mark. By and large, nuclear has been proven safe and sound.

Economics of nuclear power

Historically, the high cost of construction and decommissioning (when a plant becomes obsolete, for any of a variety of reasons, it must be decommissioned) has made nuclear more expensive than coal and natural gas fired power plant energy. But this is going to change as more concern is being focused on greenhouse gas emitting power plants.

The Congressional Budget Office (CBO) determined that nuclear power costs around $72 per megawatt-hour of output, compared to around $55 for coal and $52 for natural gas. But the numbers reflect the high construction costs; once nuclear is on line, the raw fuel costs are much lower than fossil fuel plants. So the cost of nuclear is largely a function of the construction costs, and there are a whole host of regulatory and environmental hurdles that add tremendously to this cost. The relative regulatory burden on coal fired plants is much less, and this may not be a fair accounting if the cost of carbon pollution is included in the equation.

Nuclear reactors cost anywhere between $5 to $10 billion to build, and because of the regulatory burdens can take over 15 years to complete. It is estimated that around a third of the cost of building a nuclear plant can be attributed to the regulatory mandates, and so there is considerable room to alter the economics of plant construction, since plant construction comprises the largest component of the per watt cost of nuclear power.

The British government did a study wherein they concluded that the cost of nuclear power would be equivalent to that of coal and natural gas, if the regulatory burdens were made reasonable (enough regulation to ensure safety — no more, no less). And if carbon emissions were taxed, nuclear would be less costly than coal and natural gas fired electricity.

It should be feasible to reduce the costs of nuclear construction significantly without comprising safety. Under the Nuclear Power 2010 program, the U.S. government has offered a very nonbureaucratic model for licensing, based on the French system which has proven to work very well (imagine that!). The U.S. government has even offered to subsidize up to 50 percent of the costs for the first six reactors built under this program. This is an overt acknowledgement that the government believes in a nuclear future.

Nuclear power plants have never been built in a political ambient without a lot of resistance, and this increases bureaucratic costs to a large degree. In fact, that is the goal of nuclear opponents; to make the economics unviable. If the anti-nuclear movement stopped resisting new construction so much, the costs would come down by up to 25 percent, according to some government estimates.

In the U.S., plans have been announced for the construction of 30 new nuclear plants that would provide power to 32 million homes. The per watt price of power is calculated to be equivalent to that of coal powered plants, assuming the regulatory burdens are reduced.

If the nuclear power opponents, who are the original inventors of cap and trade and other policies designed to penalize fossil fuel combustion, would do their economic calculations taking carbon pollution into account, then nuclear power would be extremely competitive, and may even cost less than fossil combustion power. The Congressional Budget Office (CBO) calculated that if the cost of carbon emissions was factored into the equation, nuclear would be the most inexpensive and safe source of electricity. It would also provide the most consistency over time, in light of the fact that fossil fuel supplies are so volatile.

In Chapter 15, I describe hydrogen fuel cells, and nuclear reactors could play an important role in producing the hydrogen needed for those cells. The entire process would be sustainable and environmentally pristine. So there are unseen economic benefits of developing more nuclear power, not to mention environmental advantages.

Chapter 9

Harnessing the Sun with Solar Power

In This Chapter

▶ Turning light to energy

▶ Battling the challenges to and economics of widespread solar power use

▶ Looking through present and future solar power options

*T*he sun sends over 35,000 times more energy to earth than humans use in all of their energy consumption endeavors. To put it another way, people use less energy in 27 years than the earth receives from the sun in a single day. That's a lot of solar energy — for all practical purposes it's an infinite supply, and renewable to boot. Best of all, it's the cleanest source of energy the world has ever known. The problem, of course, is turning all those free-for-all photons into useable energy, or ordered, energy. While humankind has been inventing creative ways to do just that for thousands of years, solar still affords a very limited applicability in the grand scheme of things. Only 0.03 percent of the world's ordered energy production comes from solar, and even though the industry is growing by over 33 percent per year, there's a long road ahead.

When solar works well — when the conditions for harvesting sunlight into ordered energy are optimal — it's one of the best alternative energy sources. Solar's effectiveness in the grand scheme of things is not doubted, but even with phenomenal growth it will never play more than a supporting role in the alternative energy scenery. But supporting roles can still win Academy Awards, and solar is one of the top nominees.

Turning Sunlight into Energy: The Basics of Solar Technology

Of all the energy sources we humans use, solar is easily the oldest (preceding even fire). There are two basic uses of solar energy (insofar as it relates to ordered energy production) — heat and generating electricity. The technologies for each are quite different, in practice, but the basic physics is essentially the same: Sunlight is captured and converted into either useable heat or useable electrical power. In the next sections I describe how light is captured and converted to useful work, and the problems and inefficiencies inherent in the process.

Understanding light

Light is all around us; most of us it we can't even see. Light is composed of individual photons, each with a wavelength and an energy, and all travelling at the same velocity.

The photon's wavelength determines the color, in the portion of the spectrum that we can actually see (much of the light spectrum is invisible to the human eye). The photon's frequency determines the energy of a photon. The higher the frequency, the higher the energy. You can visualize this by picturing two snakes, one large and one small, crossing a hot road. The large snake wriggles across the road with much longer strides (wavelength) than the small snake. To wriggle fast enough to keep up with the large snake, the small snake has to expend much more energy.

As mentioned previously, only a small portion of the spectrum is visible. At the low end is blue light, with a lot of energy (high frequency), and at the upper end is red light, which is lower in energy. Above the red zone is infrared, which is basically heat radiation. This is relevant because solar collectors are selective in which wavelengths they work with. An efficient collector will convert all the wavelengths in the spectrum into useable energy, while an inefficient collector will favor infrared, for instance. Solar water heaters convert much more of the spectrum to useable energy than PV cells, although technical advances in PV cells have enabled a much broader range of selectivity, making them more efficient.

On the passive front, windows are available which selectively filter both ultraviolet light and infrared light. Ultraviolet is damaging to fabrics and carpets, while infrared light is responsible for a lot of heat. Neither of these spectral regions are visible, and so these high-tech windows do not alter the view to the outside.

A brief history of solar power

Since time immemorial, humans have been using the sun for warmth, and devising inventive schemes to heat water and make electricity. In 1860, Auguste Mouchout, a French mathematician, invented the first solar-powered steam engine. It didn't offer much power, but it could get the job done, albeit slowly (unfortunately for Mouchout's wallet, humans don't like slow). In 1870, American John Ericsson devised a solar water trough that focused concentrated radiation onto a liquid — water or oil — and the steam pressure was used to spin turbines. As with Mouchout's invention, the job could be done, but only slowly, and only when the sun was shining. Solar engines inevitably gave way to fossil fuel power plants.

Most of the early advances in solar power were geared toward thermal applications, where the heat of the sun converted liquids into steam, which was used in steam engines. Some of the more outrageous inventions included huge mirrors, which focused a lot of energy onto a small spot size. These inventions did not fare well in windy conditions, as one might imagine. And engines like this are fixed into place, so they were never used in transport.

It was not until 1905 that Albert Einstein theorized the PV (photovoltaic) effect, which exploits the dual nature of light (photons act as both radiation and matter — a rather counterintuitive fact, but fact it is). From Einstein's work evolved the modern PV cell, which converts light energy into electrical energy, a revolutionary technological advancement that underpins the entire PV solar industry today. Einstein established some of the basics of nuclear power as well; his influence is inestimable.

Of all the solar energy that reaches the planet earth, here's where it goes:

- 35 percent is reflected away from the earth, back into space
- 43 percent is absorbed as heat radiation, both ground based and atmospheric
- 22 percent evaporates water, creating rain and water distribution
- 0.2 percent creates wind energy, or kinetic energy of the atmosphere
- 0.02 percent is used up in photosynthesis by plants

The PV cell: Generating electricity from sunlight

So how do we harness sunlight to generate electricity? With the modern PV cell, a revolutionary technological advancement that underpins the entire PV solar industry today. PV cells convert light energy into electrical energy. All solar PV systems, regardless of size or complexity or manufacturer, rely on a very simple process called the *photovoltaic effect,* which takes advantage of the fact that light photons act as both radiation and matter.

Here's how it works: Incident radiation (a fancy term for sunshine) hits the PV cell and the technology within the cell transforms the light energy (explained in the preceding section) into a raw electrical signal. Wires transmit this signal down to an inverter, which converts the power to useable electrical power.

The driving force behind the solar PV market is the underlying technology of semiconductors. Semiconductors are made of very carefully purified silicon (and other elements, when appropriate) which forms a crystalline structure with specific electrical properties. Microprocessors are made of semiconductors, as are transistors and most modern electronic devices. When a photon of light strikes a semiconductor, an electron is freed up from the crystalline structure, and this is what causes electricity to flow. All electrical flow consists of electron movement. The material that absorbs photons in a PV cell is principally silicon, which is basically sand, the same kind you find at the beach. In a semiconductor, the sand is highly ordered, whereas at the beach the sand is just plain sand.

To achieve the maximum efficiency, you need a uniform crystalline structure over large areas, but creating such a structure is difficult and expensive (although the costs are coming down with increases in production capacity and new technologies). The two current options include:

- **Monocrystalline silicon** (made of crystals of one type only) is the best quality material for PV applications because the efficiency is very high. The cost is also high.

- **Polycrystalline silicon** has a lower efficiency, but the cost is also much lower. The tradeoff, in practical terms, is that polycrystalline solar panels take up more roof space (or wherever the system is mounted) but cost less.

In the process of manufacturing a solar cell from either monocrystalline or polycrystalline silicon, small amounts of particular types of impurities (N-type and P-type) are introduced into the material in a very carefully controlled manner. N-type impurities provide for an excess of electrons, and P-type impurities leave a shortage of electrons. Either of these can create electrical flow.

The magic occurs when an N-type substrate is appropriately melded with a P-type substrate and a thin layer of silicon oxide is positioned between the two different substrates so that electricity does not flow between the substrates until light is present (see Figure 9-1). But when light *is* present, electricity flows between the substrates, and for all practical purposes the result is very similar to a battery in the way power is provided.

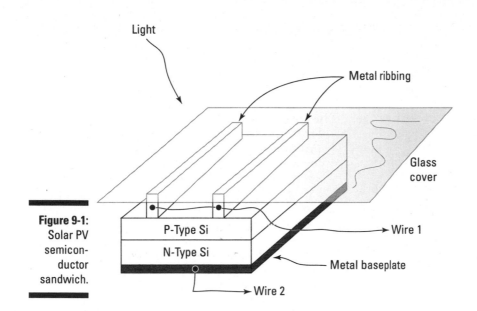

Light

Metal ribbing

Glass cover

Figure 9-1:
Solar PV semiconductor sandwich.

P-Type Si

N-Type Si

Wire 1

Metal baseplate

Wire 2

Solar PV panels, of course, aren't the only technologies that use semiconductors. You can find the same type of materials used in computer chips and other electronics. Whereas the computer industry uses small quantities of very highly purified and ordered semiconductors, the PV industry is concerned with manufacturing large surface areas as cheaply as possible, so that the collector cross section is as large as possible.

For the most part, solar PV is simple stuff and the same principles apply to large and small scale solar power systems. A large-scale system, one that powers an entire office building, is simply a huge matrix of individual elements — big arrays of PV panels are configured to produce the most power possible in the most efficient way — but it works the same way a small system (like one you'd have in your home, for example) does.

Current and future efficiency ratings of solar panels

Overall, the typical efficiency rating for a solar panel is around 16 percent, which means that of all the sunlight energy that is collected by a PV panel, only 16 percent of that energy is available as electrical power. Because there is a maximum amount of sunlight power of around 1 kW per square meter, a square meter of PV panel can only put out around 0.18 kW. There have been increases in this value, but the costs are prohibitive.

Calculating solar radiation

At sea level, on a clear day, around 1 kW of solar energy is incident on a one square meter surface (to put this in perspective, a pool pump uses around 1.5 kW's of power, and this is a lot of power). In the mountains, where the air is thinner, even more radiation is measured because much more of it gets through the thin atmosphere. However, as we all know, the sun's intensity varies over the course of a day, so it's of interest to calculate just how much solar radiation falls on a one square meter surface over the course of a day.

✔ In the summer, one can expect around 10 kWh's of energy to fall onto the one square meter surface, on a clear, cool day.

✔ In the winter, expect around 4 or 5 kWh's, under the best circumstances.

✔ When the weather is partly cloudy, in the summertime, the typical yield is around 6 kWh's.

✔ In the worst conditions, a cloudy, winter day, only around 1 kWh or less will fall onto the surface. This is why it's so cold in the winter (I probably didn't have to tell you that, did I?).

Monocrystalline solar cells have produced efficiencies of over 25 percent, while polycrystallines have achieved 20 percent. Researchers expect efficiencies will reach 30 percent, but not much higher due to the inherent physics of current technologies. This puts an upper limit on the amount of power a solar PV system will ever be able to produce on any given roof.

Efficiency may or may not be a problem. For instance, if there is unlimited surface area for installing panels (ground mounts in the middle of the desert) efficiency is of secondary concern to overall cost. In fact, the technologies that are being developed the fastest at this point put a premium on low cost, at the expense of efficiency. A new type of PV cell is being developed which is a wide, thin film that can simply be placed over an entire roof. The material, in fact, is the roof covering, and every square inch of the roof is producing power (at an efficiency of only 12 percent, maximum). Contrast this with fixed, rectangular panels that can only cover a portion of a roof.

Challenges Facing Solar Power

Solar offers many advantages: It's the only energy source that can boast zero raw fuel costs and virtually unlimited supply with no supply issues such as transport, storage, taxes, etc. It's available everywhere, even on the moon, to varying degrees. And it may be the only energy source that generates zero pollution (at least from the production, or consumption, side — invested energy is still required). In fact, from a pollution standpoint, solar power is the most energy-efficient investment that can be made, hands down.

So why don't we have solar farms pumping out electricity all over the country? Because there are limitations inherent in current solar technologies, as the following sections explain.

Technological limitations

Semiconductors are not particularly effective in creating an electrical current because their electrons are fairly stable. Therefore, a metal grid of some kind is required to overlay the PV semiconductor material.

While the grid is necessary (at least with current technology), it has the further effect of covering up some of the semiconductor, making for lower efficiencies. Plus, the process of applying the conductive grids is expensive and difficult.

New grid materials are being developed which do not shade the semiconductor materials, and this results in better efficiency. The most promising candidate is tin oxide (SnO_2).

Another issue is how much of the light spectrum current PV technology can actually convert into electrical signals.

Depending on the composition of the semiconductor, only certain wavelengths of light will cause electron/hole generation. Some photons simply don't have enough energy to create electrons or holes, and some photons have too much energy and simply bypass the entire semiconductor structure.

The spectral response of materials is also of interest, because it is important to not only collect the most amount of sunlight, but turn it into electricity. New technologies are allowing the semiconductors to react to a wider spectrum, and this increases efficiency.

Weather and temperature considerations

One of the biggest technical hurdles with all PV systems, regardless of the materials or configuration, is weather exposure. Panels are subjected to wind, dust, water, high and low temperatures, and corrosives like acid rain. Constant bombardment by solar radiation is desired, but it's also very hard on materials at the same time. Lifetime issues used to be overbearing, but recent technical improvements have assured lifetimes of over 30 years for solar PV panels.

As with all semiconductors, thin film PV substrates work better at cold temperatures — which isn't what you'd expect given that the more sunshine that hits a semiconductor, the hotter it gets. Yet, installed PV systems often output more power on a cool, breezy spring day than a hot, sunshine-intense summer day.

New designs which are becoming more common combine solar PV with solar hot water in such a way that the heat is removed from the solar PV panels. Two functions in one, and the performance of each is enhanced in the process.

The variability of sunlight

The final piece of the puzzle is *collector cross section*. Imagine a sheet of paper that you look at from different perspectives. If you set it on a table and look down, directly at it, it will look like a sheet of paper, nice and rectangular; you see the entire extent of the paper. If you look at it from the side, you will hardly see anything at all except a thin line. As you move your perspective around, the sheet grows bigger and smaller, and the rectangle becomes a parallelogram with odd angles.

A fixed solar collector works on the same principle. The more surface area that is exposed to direct sunlight, the more output the collector is capable of.

The trick to getting the maximum benefit from solar panels is to ensure that they are oriented appropriately for the region they're in.

Obviously, the number of hours of sunlight a day an area receives is important. The map in Figure 9-2 shows the average number of hours per day of sunshine in the United States and Canada. You can see that the Southwest gets the most sunshine per day (depending on the season, anywhere from 5.3 to 8.0 hours a day) — and that Canada and the northern states get the least (1.8 to 5.5 hours a day).

The hours of sunlight, however, isn't the only key factor. The quality of the sunlight, the angle at which sunlight hits a particular region, and how much atmosphere sunlight has to pass through also play a part, as the following sections explain.

Cloud cover and smog

Sunlight changes along with the weather. In cloudy regions, there is still a lot of solar energy, but it's generally diffused (spread out and unfocussed). (For this reason, solar collector panel orientation isn't so critical because light will be coming in at many different angles rather than just directly overhead from the sun.)

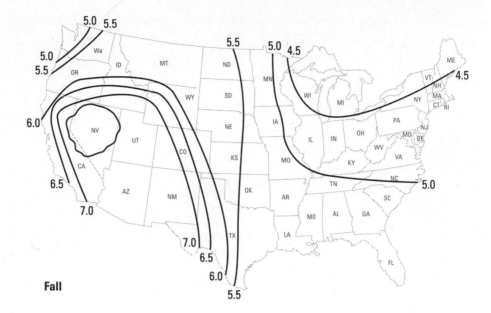

Fall

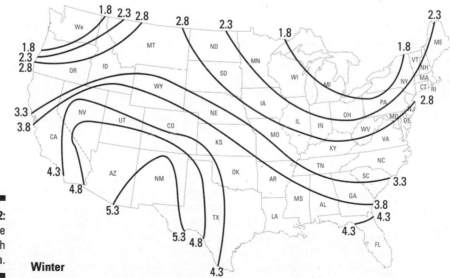

Figure 9-2:
Sunshine
in North
America.

Winter

Air pollution and smog also affect the amount of sunlight on a collector. The effect is similar to that of clouds, but depending on the composition of the smog, a lot of spectral filtering can also occur and this further decreases the efficiency. Some types of air pollution have the nasty tendency to corrode solar panels, the same way they corrode human lungs.

The position of the sun: Plotting sun charts

The more direct the sunlight on a solar collector, the more energy that collector will gather for converting into useable energy. This is a rather intuitive concept, but it's worth exploring in more detail.

The position of the sun may be plotted with two angles (*azimuth,* which is the angle from true south, and *altitude,* which is the angle from level, or in most cases the horizon), as shown in Figure 9-3.

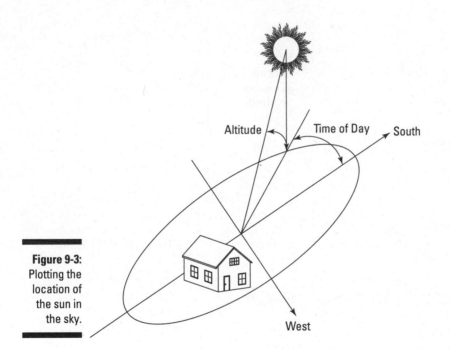

Figure 9-3:
Plotting the location of the sun in the sky.

Figure 9-4 shows a graph of the movement of the sun over the course of a day. The arc in the middle represents either spring or fall. All other paths lie somewhere between the two extremes, represented by summer and winter equinox, which are the longest and shortest days of the year. By using computer analysis programs, it's possible to predict the amount of total sunshine a collector will receive on any given day of the year. In the winter much less sunshine will be available, and this is indicated by the area under the winter curve. In the summer, the area under the curve gives the expected sunshine. System productivity is always of concern, because it determines the payback on a solar investment, so a good estimation of the total sunlight will yield a good estimate of the total expected system output.

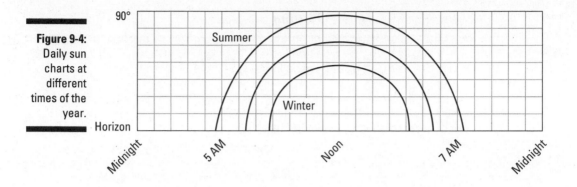

Figure 9-4:
Daily sun
charts at
different
times of the
year.

When the sun goes behind a mountain or a tall building, there will be no direct sunlight at all. If the sun goes behind a tree, there may be some direct sunlight, but mostly there will be shading effects. Also, the horizon changes the time of dawn and dusk.

Sunlight intensity

In addition to how much direct sunlight a collector gets, sunlight intensity is important. When the sun is lower in the sky, solar radiation must pass through more atmosphere, and it's therefore reduced by scattering and absorption. For this reason, solar exposure is better in the mountains than near sea level simply because the air is thinner and scatters less sunlight. (That's why sunburns on the ski slopes are much worse than the kind you get at the pool.)

Getting enough power

Current technologies for PV systems (as well as water heating systems) capture only 30 watts per square meter, on a round the clock average. To put this in perspective, New York City consumes 55 watts per square meter of energy on the same round the clock average. So to power New York City, an area of twice the city would have to be covered with solar collectors (and beneath this would be no direct sunshine, which might not make for such a benign environment). A uranium mine produces a million times more energy, per surface area.

Using current technology, you'd have to cover a huge surface area of the earth to generate enough solar power to put an appreciable dent in the world's energy demand. Even if that were possible, it would have a severe impact on the ecological balance (think of the poor sunshine starved plants beneath the collectors).

The point is, solar will never be more than a sidebar in the overall energy picture simply due to the logistics of covering expansive areas with panels.

Earth: Going full tilt

The Earth is tilted on its axis 23.5°, and the following figure shows what happens to the position of the sun in the sky over the course of a year.

Regardless of where you live, the difference between the sun's peak angles in the sky from December to June is 46°. Regions closest to the poles experience seasons when the sun never shines, and six months later, seasons when the sun never goes below the horizon (known as white night).

The optimum elevation angle for a solar system depends on latitude. In general, the optimum tilt angle is equal to latitude, for this will ensure the maximum amount of sunlight exposure over the course of a year. Even better would be to manually change the elevation of the solar collectors periodically over the course of a year to "follow" the sun's elevation in the sky.

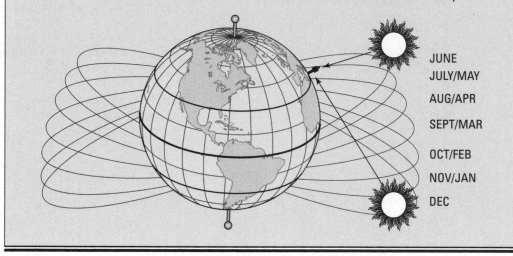

JUNE
JULY/MAY
AUG/APR
SEPT/MAR
OCT/FEB
NOV/JAN
DEC

The Economics of Solar Power

Although a number of studies have shown a net positive impact of solar energy, particularly in communities dedicated to large-scale development of solar power, the most basic reality of solar power is that it is viable only in sunshine rich climates, particularly the Southwest and Pacific coastal areas. Even with government subsidies, solar is not now, nor will it likely ever be, viable in cloudy, rainy climates. However, if the price of solar decreases enough, this may change, although the vehicle of change will be investment capital, and when solar lags so far behind other alternative energy sources, investment capital also lags behind.

The following sections provide you with info on the economical impacts of large-scale solar power use, including the benefits as well as the economical challenges solar power faces.

In 2005, the U.S. economy consumed 100 quadrillion Btus of energy, of which less than 1 percent was solar. The impact that solar makes on the entire renewable (alternative) energy industry is very small. Furthermore the consumption of all energy sources in the last ten years has increased 25 percent, but solar production has only increased 17 percent, so solar is not even keeping up with the economy at large. This is due solely to the peculiar economics of solar, not the available technologies.

The dollars and sense of solar power

As you can see from this list, not all of the economic benefits of solar power can be enumerated in monetary terms — you can't put a price on good sense:

- Solar communities are cleaner, with less carbon footprint. The benefits of this are intangible, but nevertheless people are aware and there is a decidedly increased community spirit.

- Solar communities tend to support recycling programs and other environmental improvements.

- In communities dedicated to solar, much less of the community's money is going out of the community.

- With solar energy, local jobs are created, and the jobs are stable and sustainable and skilled, meaning high pay and good benefits.

Yet, despite these positives, the economics of solar power are not as good as for conventional energy sources (combustion sources): upfront investment costs are high; payback may take a long time; and the investment is illiquid and fixed. Most people do not want, or cannot, lay out a big investment solely for the purpose of rationalizing their energy consumption.

Looking at solar adoption, sector by sector

Traditionally, there have been two main sectors for solar: residential and commercial. Residential customers are more apt to invest in solar due to "green" sentiments than commercial customers, so the economics are skewed in favor of residential. In 2002, solar was 13 percent of all renewable energy consumed in residences but only 6 percent of commercial consumption.

It's an open question just how much the green factor plays in payback calculations, but it is a fact that many residential solar customers are willing to pay a premium to eliminate their personal carbon footprints. This trend is increasing, with all the news of global warming and society's general sense that we each need to chip in to help solve the problem.

More recently, independent power producers are also going to solar, mainly because of government mandates that power producers must wean themselves off of fossil fuel energy sources. From 1989 to 2003, independent power producers increased their solar production by 67 percent, which is much higher than the overall power production increase. So independents are investing more and more into solar, but not because of economics — it's because of mandates.

Commercial customers of solar are generally concerned with the bottom line on their business, although businesses that install solar can advertise that they're "green," and in many markets that makes a favorable difference with customers. For instance, a landscaping company that installs solar panels on their roof, and advertises that they are a "green" company, can see their sales increase. This has the effect of making the payback of the solar investment much better. It should be added that this effect can be abused; some commercial entities install a minimally sized system just so they can advertise their greenness; they're not really green, but they're claiming to be. Perhaps someday the government will regulate what standards a business must meet before proclaiming itself "green."

Looking at infrastructure

The main economic factor for solar PV, as well as solar hot water, is the lack of supporting infrastructure. Fossil fuels drive the vast majority of the world's energy market, and as such there is a tremendous infrastructure dedicated toward that energy source. Solar, however, is a relatively isolated technology, and contractors often have a difficult time maintaining their businesses in the light of varying government involvement. In particular, the worldwide supply of PV panels fluctuates quite a bit, and this results in price disparities and inconsistencies.

Better solar infrastructure would include more collector manufacturing plants. The majority of those in existence now are overseas, so the government should subsidize American solar equipment manufacturers. In fact, most solar equipment is manufactured overseas, and this causes bottlenecks and uncontrollable price fluctuations. Above all, the government needs to be more consistent with its subsidies and tax breaks.

Another important economic question concerns development costs for new technologies, and jobs that will be created in the process. In the burgeoning green paradigm, sustainable jobs are just as important as sustainable energy sources, and so the development of a formidable solar industry is good for society as a whole. Solar is produced locally, and so the jobs and infrastructure are local.

Comparing cost: Solar versus fossil fuels

As with any alternative energy source, the question is how does the cost of solar power compare to conventional fossil fuels? Current electrical rates around North America range from a low of around six cents per kWh to over 50 cents, particularly in tiered rate structures where a progressive price is applied to the quantity of energy consumption used in a building. The more consumption, the higher the kWh price goes. A rule of thumb in the industry is that solar PV costs around 15 cents per kWh. This includes up-front investment as well as maintenance and lifetime costs. Plus, this assumes best case scenarios, namely a lot of sunshine and good weather. So the fact is, solar PV is not economically competitive with grid power except under certain conditions.

However, if pollution mitigation is taken into account, the economics change quite a bit. Solar power is basically pollution free, aside from invested energy, and that's improving every year. If cap and trade policies are instituted the net effect will be that solar becomes much more competitive. In fact, of all the alternative energy scenarios, solar will gain the most from a *carbon tax,* a tax placed on carbon outputting energy sources — for instance, a 50-cent surcharge on a gallon of gasoline due to its high carbon output.

The economics of net metering

When the solar system is outputting more energy than the host building is using, excess power is available. Net metering (also referred to as *intertie,* or grid connect) means that you can sell your excess power back to the utility company. Net metering is great for PV system economics because every solar system, whether PV, water heating, or another type, has a certain maximum capacity. While maximum capacity depends on a number of things — on a cloudy day, system capacity is much less than on a sunny day, for example — consumption is governed simply by what you need and want. In a typical household, power consumption peaks in the morning and evening while solar production peaks in the middle of the day, when the sun is highest in the sky. With a PV system tied in to the grid, you can use as much power as you need, and your PV system merely contributes whenever it can. What you don't use from your solar system is all sold back to the utility.

Without net metering, or intertie, solar PV economics would be hopeless under most circumstances. In fact, solar PV has proliferated solely due to net metering laws.

The government has mandated that utilities not only allow solar PV customers to connect to the grid, but that they pay the same amount for power they receive from solar systems as for the power they dispense to their customers. Without the mandates, the utilities would probably not allow solar at all. In fact, without mandates, the utilities would probably demand the first born of every one of their customers.

In some regions, solar PV is even more advantageous because the utilities offer time of use (TOU) rate scheduling, where electricity rates are highest during peak usage time (generally, from noon until 6 p.m.). The timing is perfect for solar PV systems because they generate a majority of their power during peak time (remember, by law the utilities must pay you the same rate they charge for power, so at peak time they must pay you peak rates). Then later, when you use more power than you're generating, you get power back at a much smaller price. In this way, the size of a solar PV system can be leveraged to increase return on investment.

Government Incentives for Solar Power

The upfront investment cost of solar power is very high. Most people don't have upwards of $20,000 to spend on solar PV equipment. And those that do would often rather have a new kitchen or a swimming pool, both of which offer a much more immediate, tactile enjoyment. So without government subsidies, solar would be little more than an asterisk. Allocating accurate costs is one of the keys to making alternatives work, and solar is certainly not alone in that its costs need subsidizing. When cap and trade systems are embraced, solar will be a much more viable energy source. In fact, with cap and trade, solar will be the best alternative energy candidate of all because there is no carbon outputs at all.

The one big factor in determining how much money will go into solar development and production is government incentives. Without government incentives, the economics of solar power are non-competitive. This is why the federal governments of a number of countries provide rebates and tax incentives for the solar industry. The thinking is that if the government provides enough development money, over time the industry will catch up economically to the fossil fuel industry, and become competitive without government subsidies. So in essence, government intervention is creating a new industry, one that is beneficial and desirable for society on a number of fronts.

The government is interested in subsidizing solar PV for three main reasons:

- Solar power is the ideal solution for peak-power generating problems. Solar systems output the most amount of power in the afternoon, when the sun is shining the brightest.

- Solar PV is the cleanest, most efficient source of energy. There is literally no pollution in the operation of a solar PV system. There is some invested energy, of course, but the raw fuel costs are zero, and the operating pollution is zero.

- Solar PV is distributed, meaning it does not come from one large, centralized source. This is advantageous because it affords a less risky society, both in terms of assuring continuous power availability and inoculation from terrorist acts.

In countries such as Japan and Germany, generous solar subsidies have been in place for decades, and the infrastructure is far ahead of that in North America.

Federal tax credits

Federal tax credits are the largest solar subsidy, constituting 30 percent of a solar system's value, both for commercial and residential. One must pay taxes, however, in order to exploit the tax credits. Plus there is a need to finance the amount of the credit at the purchase point of the solar system. It may take over a year, or even more if credits are spread out over time, to recoup the original expense. Yet at 30 percent of total system cost, this tax credit is a real gift.

State incentive programs

The state of California leads the nation in solar installations, and it's largely because of a massive subsidy known as the California Solar Initiative (CSI). Up to 25 percent of the cost of a solar system is rebated to the customer, but it's even better than that. The rebate amount goes straight to the solar contractor that installs the system so the customer doesn't have to front the money prior to receiving the rebate (with most rebates, you have to buy something at the retail price, then apply for a rebate which comes later in the mail).

One man's bailout is another's bonanza

In the famous (or infamous) 2008 bailout bill passed by congress amidst great political pressure, the residential tax credits for solar power were drastically changed, to the benefit of the entire solar industry. Prior to the bill, the tax credit was 30 percent, with a cap of $2,000 on residential installations (and no cap on commercial applications, which strongly favored commercial). Under the new law, the cap on residential was removed, which resulted in a tremendous surge in residential demand.

This illustrates the precarious nature of the economics of alternative energy, in that the entire market conditions can change literally overnight. With this kind of uncertainty in the market, potential investors are wary because these kinds of changes can go both ways. A potential investor looks not only at expected returns on investment, but also at risk, and when there is a big question mark looming overhead the risk is increased.

Governments need to stabilize alternative energy markets with more consistent approaches to the subsidies that are offered. It's not enough just to step in once in awhile and make some money available. The money needs consistency as well and this is a difficult end to achieve when politics is the central player in the drama.

In addition, the state of California mandates that a solar customer cannot have his/her building reassessed to raise property taxes. With most improvements in a home, particularly the ones that require county building permits with all the associated code inspections, the county reassesses the value of the property, and the property taxes go up. Not so with solar.

Washington State has offered energy consumers 0 percent loans for the purchase of solar systems, both PV and hot water. These loans are available through the utilities so that customers may simply have solar installed, and instead of paying their utility bills, pay off the loan for the solar equipment. It's common for the loan payments to match the customer's power bill, so the net cash flow is the same. And these loans are assumable to a new owner.

Hawaii just passed a bill requiring all new homes to install solar water heaters. The cost of energy in Hawaii grew from less than 7 cents per kWh in 2003 to over 15 cents per kWh in 2005, due mostly to increased demand and limited production capacities. Hawaii is not connected to the American power grid, obviously, so their unique situation calls for unique measures.

Arguments against subsidies and tax breaks

Obviously, when the government subsidizes any form of alternative energy, it's the taxpayers who are footing the bill. So society, in general, is subsidizing those customers who install solar. This is seen as unfair in some circles, since it favors the wealthier individuals who can afford the up-front cost of solar equipment. Just because the payback economics of a solar investment are favorable doesn't mean that people will install solar because a majority of the population are living from month to month paychecks and don't have the means for a large investment.

Another argument against solar is that subsidies are only available to those who have homes and businesses in sunny climates. Even in sunny climates, homes with a lot of surrounding trees won't be able to exploit solar power, so in some ways solar subsidies are anti-tree (as opposed to tree hugging). And many roofs are not oriented well to exploit solar – roofs at a slope of 25 degrees, facing true south, afford the best production. Some roofs are so hopelessly off the optimum that a solar investment is simply not worth the cost, even with all the rebates and tax credits. And roofs that have many separate facets are also not amenable to solar because it's impossible to achieve enough uninterrupted surface area for the collectors. At the very least, luck plays a role in the equation because many homes and businesses simply don't have the requisite conditions for productive solar.

Looking at the Full Range of Practical Solar Options

The range of different solar applications is far broader than any other type of alternative energy sources, and this enables solar to find its way into more niches. The following sections outline some general ways that solar power can be used.

Passive solar

Passive solar objectives are straightforward: allow sunshine into the interior spaces of a building when heat and light is needed, keep the sunshine out during hot summer months, and manage optimally the heat that is allowed to enter.

Passive building designs allow for not only solar lighting but solar heating and cooling systems. Passive solar works best when designed into a building's original structure. Retrofitting a building into a passive solar masterpiece is much more difficult, and the economics can be a stretch.

These objectives are accomplished by the following:

- **Proper orientation of the building.** Orientation is best when the long side of a building is facing true south to expose the most surface area to the sun so that sunlight can be channeled into the rooms where the family spends the most time. Great rooms, family rooms, living rooms, and kitchens should always be located on the southern exposures, and should exploit the solar design elements. This ensures a more active interaction between the natural elements and the building dwellers.

- **Strategic placement of windows, skylights, and awnings, as well as deciduous trees and trellises on the outside.** From an environmental standpoint, the strategic placement of trees is the best possible solution because buildings shaded appropriately with trees can see up to 20 percent cost reductions in heating and cooling costs.

- **Using thermal mass to store heat in the winter.** Thermal mass acts as a ballast to maintain a consistent temperature in the building, regardless of season. Concrete floors have a lot of thermal mass, as do tile and bricks and stones (water does as well). Wood and carpet have little thermal mass.

- **Making optimum use of insulation and ventilation.** Regardless of whether one is trying to heat or cool, the ability to control heat movement is important. Ventilation is also important because it provides fresh air, and also allows for moving heat, as appropriate. Natural ventilation schemes such as the chimney affect and thermosiphoning are always preferable to machine driven, active means.

- **Using light tubes and skylights enhances the natural feel of a building's interior, and reduces the burden on electrical lighting.** A good solar building will require no active lights during a typical work day.

Buildings that incorporate passive solar features offer energy bills up to 70 percent lower than conventional buildings. Plus the interiors are more naturally lit, which lends a sense of spaciousness and makes people feel closer to Mother Nature. Many localities now require commercial buildings to be designed using passive solar as much as possible, and since it doesn't cost much more to include passive solar the building industry is readily complying.

Solar PV

In one fell swoop a solar PV system can completely offset an entire electric utility bill while generating zero pollution. Millions of photovoltaic (PV) systems are being installed all over the world, and the costs are decreasing due to economies of scale and technological improvements.

Solar PV (photovoltaic, or the generation of electric power from the sun) is coming into its own. The power generation is local (thereby cutting out the inefficiencies of long grid transmission lines and ancillary support equipment) and completely free of pollution. Following are some ways that solar PV technology is being used today:

✔ To run pumps for irrigation, remote cabins, livestock watering applications, etc. These systems are usually well away from electrical utility hookups.

✔ To power roadway signs, remote gates, and other applications where utility power is not available.

✔ To power remote applications such as roadside call boxes, illuminated highway signs, navigational systems, buoys, unmanned installations, gate openers and electric fences, and remote pumping

✔ To directly power DC electrical motors. If you don't have access to grid electricity, this can be very useful and economical. If you simply want to cut back on pollution, this is also a good way to make a contribution. This type of system works very well for remote cabins, and where electrical grid power is not available. For instance, ranchers use such systems to provide water to cattle located in the boondocks.

Large scale solar concentrators are being developed which are capable of turning turbines and creating commercial amounts of electricity. At present, these systems are little more than research novelties but as with solar PV, the situation is rapidly improving.

Solar hot water systems

The complexity of solar hot water systems varies, but they all do pretty much the same thing: heat water. The vast majority of hot water systems are used in third world countries, and are made of very simple materials.

A solar hot water heater can be made out of little more than a big metal barrel painted black. Black absorbs the solar radiation, which then conducts into the water, which can then be used for showers and baths and cleaning. In simpler systems, the temperature is not very well regulated, but these systems are generally used where there is no electrical power at all, and they are considered luxuries. More sophisticated systems use collectors made of black coated copper, and have sealed chambers with very good insulation properties. These systems heat water very efficiently.

A distinct technical advantage of solar water heating technology is that the storage of the collected heat energy is straightforward – all you need is a suitable tank. Some collectors, like the black-painted metal barrel mentioned earlier, both collect and store the heat. More sophisticated systems collect the heat then channel it into a special storage chamber where it stays until needed.

There are other advantages as well: Most solar water heaters require little or no external power, there is no pollution generated with a solar water heater, and the systems are virtually noise free.

Solar water purification systems

Solar water purification systems provide remote villages in third world countries with a steady supply of clean, pure water for drinking and cooking. Other options don't even compete in terms of cost and reliability.

A solar water purifier is very simple, in operation. A space is enclosed by insulation on the bottom and sides, and a sunlight transmitting window is located on top. A water filled trough is located at the bottom. When the sun shines through the window it heats the interior of the chamber, and particularly the water in the trough, which evaporates at a speed depending on the interior temperature, which is a function of how much sunlight enters the space. This vaporized water is free of impurities (much like a still works to separate components of a complex liquid) and condenses on the window and sides of the enclosure (this is the same effect that causes water to condense on a cold glass). A suitable collecting trough gathers the purified water and delivers it to a container on the outside of the enclosure.

Utility-scale photovoltaic (PV) plants

Utility-scale PV plants are large arrays of photovoltaic systems set up by the utilities. A combined 1 MW of peak power is now being generated by these plants worldwide, with Germany leading in capacity, followed by the U.S. The largest system of all is located in Prescott, Arizona, with a capacity of 5 MW peak.

Costs for these plants are coming in at around $5 per watt, but some plants are more expensive, and produce less net annual production. The economics are not competitive with fossil fuels, but because of government mandates stating that utilities must produce a certain minimum amount of their total power with renewable resources, these large plants fulfill an increasing need. With decreases in the cost per watt for solar panels, these plants will become much more common.

Large-scale solar condenser power generators

Large-scale concentrated solar systems use arrays of individually focusing mirrors, aimed at either a single point or a line. The concentrated heat can be quite intense, depending on the surface area of the mirrors, and the quality of the focus. The high heat is used to vaporize some form of liquid, and the pressure is used to drive a turbine, which in turn creates electricity. Mechanical trackers continuously focus the sunlight onto the solar collector or collectors.

Southern California Edison built a 10MW facility in the Mojave desert. It cost $140 million, which translates into $14,000 per kW. Fossil fueled plants, in contrast, cost around $1,000 per kW, so the economics are not good at this point. But since the fuel is free, the long term economics are more favorable when lifetime considerations are added to the mix.

Perhaps the greatest advantage of these systems is that they produce a majority of their power during the middle of the afternoon, and this helps utilities with their peak-power emergencies. Peak-power consumption occurs on hot summer days when everybody's air conditioners are running full bore. The utilities cannot keep up with demand, and small-sized backup power generators are required. Large scale solar thermal power generators are ideal backup generators because they output the most power when the sun is highest and hottest, and this is exactly when the most air conditioners are being run.

These types of systems are generally built in deserts, where sunshine is prevalent and where they're removed from population centers.

As with everything, there are a few downsides to these systems:

- ✔ Whenever mechanical parts are in play, reliability suffers, and costs rise accordingly. Plus there will be a requirement to have more on-hand labor force to deal with the problems.

- ✔ Building a system requires a tremendous amount of capital.

- ✔ As with all solar systems, the system output tracks the sunshine availability, so there is a random element at play.

- ✔ Ultimately, there is a limit to how large a system can be built. In order to generate great quantities of power, a good number of individual plants would have to be built.

Future Prognosis of PV

Solar energy holds a great deal of promise, but in order for that promise to be realized on a bigger scale than it is today, the cost issues must be addressed so that more people can take advantage of the technology. Widespread use of PV will depend on new financing schemes, the best of which are explained in the following sections.

Financing through utility companies

Utilities finance solar PV directly with their customers, and in the process the customer pays the utility for the loan, as opposed to buying energy from the utility. In this way, the customer builds up value (equity in the home) as opposed to simply paying each month for grid power.

The customer's suitability for PV will still be a large factor, as some homes simply don't have enough sunshine to make solar work. So the politics of mandating this become murky because only those who are lucky enough to have suitable homes will qualify for the benefits. Politicians make more people happy when luck is not a factor in the equation. And ultimately, politicians run the utilities, which are publically regulated.

Leasing schemes

Leasing schemes for PV systems are beginning to proliferate. A company installs a system, and you pay that company an agreed upon lease amount. You then own all the production of the system. These deals are very similar to a lease for an auto, with residual value, maintenance responsibilities, and so on. The leasing company still owns the system, so if you default on your payments they can repossess it.

This works well for solar because the lease payments are usually less than the value of the electrical energy that the system produces. Plus, you don't have to lay out a big wad of up-front cash in order to capitalize on solar power.

Buying into solar farms

Picture this: A company buys very cheap land in the middle of Nevada, where the sun shines nearly every day of the year. They sell you a solar PV system and they install it not on your roof or your property, but in a designated plot somewhere on this very productive land. They can sell you the system for much cheaper than the standard contractor price because they install a lot of systems right next to each other, and don't need to go to your home and climb around on your roof. They also get steep discounts on parts because they buy so many.

The electricity that the system generates is fed into an on-site meter, which measures the quantity and credits it to your account. This power is fed directly into the grid at the site of the solar farm.

You own all the production for the system, and the company can move the tilt angles of your panels several times a year (they move everybody's panels, so it's not difficult or expensive) so that the production is optimized. The economics of this particular scheme will dwarf any other PV finance scheme available. So why hasn't this happened? Utilities are resistant to anything that makes this much sense.

Mandating and amortizing

New buildings and homes will be mandated to have solar PV systems. In some states, like Hawaii, solar hot water systems are mandated. Then the cost of the equipment gets amortized into the cost of the home. Commercial buildings will see the same types of mandates; it's already happening with energy-efficiency equipment.

Chapter 10

Treading Water with Hydropower

In This Chapter

▶ Catching a glimpse of hydropower in the U.S. and the world

▶ Revealing hydropower systems built on river ways

▶ Examining hydropower systems that exploit tides and waves

Water is one of the earth's most important elements, and perhaps its most important natural resource. The kinetic energy of a fast moving river may be converted into electrical power that can meet the energy demands of entire regions. Water may be dammed and stored and the potential energy tapped to produce electricity. Waves and tides may also be used as energy sources. Humans have tapped the power of water for millennia, to varying degrees of success.

In this chapter I describe the various technologies used to harness water power, and I analyze the pros and cons. The bottom line, at least in the United States, is that hydropower has largely been tapped. There are more possibilities, but many dams are being torn down, and new ones are not replacing these at the same rate so hydropower is actually declining.

The Hydropower Story

Ancient civilizations utilized waterwheels that scooped up buckets of water mounted onto rotating wheels. This power fed mills and industrial machines, not to mention the pumping of the water itself to nearby urban areas. In the sections that follow, I pick up where ancient civilizations left off, taking you through a brief history of hydropower, and I walk you through the basics of hydropower systems as well as global trends in hydropower production and usage.

Hydropower history

At the dawn of the Industrial Age, waterwheels were commonly used to power mills and heavy machinery in the Northeastern United States. But the power outputs were limited by the availability of the water supply, which can fluctuate wildly with weather patterns. In some cases, entire mills were shut down during droughts and the unemployed workers could only look to the skies in anticipation.

Then came technological advancements that enabled large river damming (which allows water flow to be restricted and therefore managed and controlled), advancements in turbine technology, and massive government programs that promoted the building of large-scale dams for energy. One such project was undertaken by the Tennessee Valley Authority (TVA), created during the New Deal–era, to construct dams along the Tennessee River in an effort to control floods, improve navigation, and provide electricity to residents and industries in the Tennessee Valley. The energy derived from damming large rivers with reservoirs and powering huge turbines to create electricity is called *impoundment hydropower* (the water is "impounded").

During the last century, the demand for electricity has resulted in the damming of large rivers worldwide, particularly in the developing world. During this time period, the U.S. led the world in terms of dam building on large scales.

But today, the well is running dry. There are fewer and fewer opportunities to build dams because our major rivers are dammed already. There are ecological questions as well. Despite the fact that hydropower emits no greenhouse gases (and this is certainly laudable), ecosystems have been drastically altered, and this affects not only wildlife but human social systems. As a result, new dam construction has virtually been halted.

In 1940, hydropower provided around 40 percent of all U.S. electrical requirements. Today, hydropower only provides around 8 percent of the nation's electrical needs, and this number is consistently declining.

Global trends

Worldwide, over 2,700 Tetrawatt-hours of electrical power were produced from 750 gigawatts of hydropower capacity. However, since 1990, hydro has seen a drop in production, from 18.5 percent of the total energy output to less than 16 percent — this in spite of the increase in dam building in developing countries. Part of the problem is that fossil fuels are too inexpensive, and too abundant.

In the U.S.

Impoundment hydropower accounts for anywhere between 7 percent and 12 percent of total U.S. electrical consumption (or 40 percent of our total renewable energy consumption).

The U.S. currently produces 80,000 megawatts of hydro power. Another 18,000 megawatts is produced in pumped storage systems. While there are over 75,000 dams over six feet high in the U.S., only 3 percent are used to generate electricity. In addition to these 75,000 good sized dams on its river ways, the U.S. has thousands of smaller ones that don't produce power (what do they produce?) This does not include dams built by beavers and young boys, which may number in the millions.

In recent years, a number of dams have been torn down in the U.S. Between 1999 and 2005 over 120 dams were dismantled and there are a good number more dismantling projects underway. At present, there are very few new dams on the drawing board in the U.S. The environmental impacts have become obvious, plus there are simply no more readily available waterways suitable for economical electrical production.

The declining reliance on large-scale hydropower is a trend that's unlikely to change, as other alternative energy sources are proving more promising.

In other places around the world

Norway gets all of its electrical production from hydropower, while Brazil derives over 80 percent. These countries have rich hydropower resources, and their populations are low along the rivers used for the dams. Canada is the only industrialized nation that plans on expanding its hydropower capacities.

China's Three Gorges Dam was recently finished. The environmental impact has been staggering, not to mention the number of displaced people. The dam will output 18,200 megawatts of electricity, the equivalent of 40 million tons of coal annually, or 20 nuclear plants. The dam stretches over a mile and features a vertical drop of over 750 feet. If Mother Nature decides she doesn't like the dam, and big earthquake should do the trick.

A quick look at hydropower systems

There are a wide range of hydropower engineering schemes, each dependent on the combination of flow and head available at each point on the river.

There are two characteristics that determine how much power or energy can be obtained via hydropower: flow, and head. *Flow* is the amount of water that flows past a given point in a given time period. *Head* is the water pressure, or how hard the water wants to flow. A very deep dam has a lot of head pressure.

Hydropower basically derives from the sun. Solar radiation causes evaporation (mostly from the oceans) and this then condenses into rain, which falls over land. Streams form into rivers, and the water flows back toward the ocean due to gravity. So hydropower is actually a combination of gravitational potential energy and solar energy working in conjunction.

River systems

Traditional hydropower systems use the current of river ways to produce power. In impoundment hydropower systems (by far the most common hydropower system in use currently), the river is dammed to create a reservoir, and then the water is channeled through a turbine that produces electricity. With most impoundment dams, the entire river is dammed up so that all flow downstream must come through the dam, or the gates. This not only creates electrical power, but it serves to control the river's flow so that flooding may be controlled as well.

Less common variations of hydroelectric systems include the following:

- **Diversion systems** are basically the same as impoundment systems, but they only use a portion of the river's total potential to generate electric power. Much of the water is fed on past the dam/generator so as to keep the river as natural as possible. This is easier on the environment than an impoundment system, but much less productive.

- **Run-of-river systems** simply use the kinetic energy of a fast moving water and convert that into electrical power. The head pressure is minimal in this case, but the flow is strong. The river basically runs slower after it passes through a run-of-river system. This is generally not as environmentally damaging as impoundment and diversion dams.

- **Pumped storage systems** generate electricity during peak times, and pump the water back into a holding reservoir during off-peak times. While these systems, on net, produce negative amounts of energy, they are very good at allowing a utility to react to high peak demands, and so the economics are actually favorable since high peak-demand power comes from inefficient backup generators that cost more per kWh. Pumped storage systems can be viewed as huge batteries, capable of storing electricity.

- **Small-scale hydropower systems** generate a single home's worth of electrical power. This option is on a par with solar and wind power, so it will likely see a growth in the coming decades. Small-scale systems obviously alter the environment much less than the massive impoundment dams on big rivers. But if the trend grows too large, aggregate environmental impacts may begin to materialize.

Oceanic systems

A great deal of energy is contained in the world's oceans. Tides move in and out with great force that can be harnessed to drive turbines. Waves can be "captured" for the vast amounts of energy that move across the surfaces. Oceanic systems are not as common as river way dams at this point, but with improving technologies, there is much greater promise, and the potential has hardly been tapped.

- ✔ **Tidal power systems:** Tidal power systems (also called *tidal barrage systems)* use the power of rising and falling tides to create electrical power. When tides come in, the water is captured in a reservoir, or basin. When tides go out, a turbine spins and creates electricity. It's not much different than an impoundment dam, or a pumped storage dam where the pumping is done by the tides instead of manmade machinery.

- ✔ **Wave power systems:** Waves may also be tapped for their energy. In these systems, some variation of a float mechanism is used to pump a turbine periodically with the rising and falling motion of a wave. Waves offer tremendous amounts of energy, but the supply is inconsistent compared to a tidal barrage system.

Both of these systems are minor players in the hydropower game, but with the increasing cost of fossil fuel energy sources, research and development money is starting to flow. Both of these minor systems offer favorable economics and very low environmental risk, so they hold more promise than impoundment systems located on rivers.

Impoundment Systems — The Big Kahuna

Impoundment systems use a dam to capture and contain water in a reservoir. Gates control the flow of the water into a large pipe (called a *penstock)* that feeds the pressurized water into the turbine, which spins an electrical generator. The water is then discharged back into the river way.

The depth of the reservoir creates head pressure. The deeper the reservoir, the greater the head pressure and the greater the potential to generate power. The amount of water flowing through the turbine (in combination with the head pressure, which depends on the depth of the water in the dam) can be varied to control the amount of electrical power output of the hydropower generator.

The total potential energy that can be generated is a function of how many gallons of water are in the reservoir times the head pressure. Low depth, wide dams are not as good as high, narrow dams in terms of power output, although the total energy potential may be the same.

Impoundment systems are by far the most common type of hydropower generating system in use. A flowing river with a section amenable to building a dam is the only requirement, which explains why this type of system is found in over 160 countries worldwide. Because, according to estimates, only a third of worldwide hydropower potential has been tapped, the growth in foreign hydropower promises to be very high in the next decades.

Advantages of impoundment systems

The attractiveness of impoundment systems is understandable: Water is a renewable source of energy, and the natural rain cycle constantly replenishes the water supply, at zero cost. No fossil fuels are needed, and the supply of raw fuel is virtually automatic. Hydropower dams also reduce a nation's reliance on foreign energy sources. Following are some other key benefits:

- ✔ **Hydropower is cost competitive with all other energy sources, and in many cases it provides the cheapest energy obtainable.** As much as 90 percent of the potential energy of flowing water can be transformed into electrical energy, which makes it one of the most efficient energy sources. Fossil fuel plants can generally achieve only 50 percent efficiencies, at best.

- ✔ **Hydropower can be altered very easily in terms of immediate power output.** Hydropower plants can be easily turned up or down, at will. It's a simple matter to simply lower or raise the gates to decrease or increase the flow through the turbines. When load conditions change, hydropower plants are the perfect solution.

- ✔ **The storage is safe and clean, without the tremendous hazards associated with fossil fuel storage.** Hydropower is unique in that it can both store potential energy, then convert it to immediate use. The reservoirs can serve the dual purpose of supplying potable water to communities nearby. Plus, storing hydropower potential energy results in recreation usage, and no other energy source can advertise that advantage. In fact, the recreational opportunities that impoundment dams afford may be even more valuable than the energy they provide.

Many dams are built solely for their recreational value. In 1996 alone, over 80 million user days were spent fishing, camping, hiking, and boating above impoundment dams in the U.S.

> ✔ **Large reservoirs, a key component of impoundment systems, help ensure consistent energy output despite weather fluctuations.** During periods of drought, the water contained above the dam will slowly fall in depth, but there's generally enough water to last several months, if not years.

> ✔ **Hydropower produces no toxic wastes, pollutants, or carbon dioxide and greenhouse gases.** There is no pollution generated, in particular no greenhouse gases. There are also no sulfur emissions or other sooty waste products. There are no fuel transport lines required, other than the rivers and streams themselves, and these are certainly much better than pipelines and highways full of oil tankers.

> In 2000, U.S. hydropower plants offset the burning of 28 million gallons of oil, 121 million tons of coal, and 750 billion cubic feet of natural gas. In the process, 1.6 tons of sulfur dioxide and 1 million tons of nitrogen oxide were spared from the air. Seventy-seven million tons of carbon dioxide were also offset, and this may be the greatest advantage of hydro over fossil fuels. This savings of carbon dioxide represents the emission by 60 million passenger vehicles each year.

These benefits are both blessing and curse, as nations sometimes come to blows over water rights. For instance, a nation that's upriver can dam the river, causing undue hardship downriver. What's the solution?

Environmental impacts of impoundment dams

The story is not all a bed of roses, however. There are a range of environmental problems associated with impoundment dams, and these factors are becoming extremely important as time passes and the effects become more obvious.

Impoundment systems drastically alter a river's natural ecosystem, both above the dam and below. Not only is the river affected, but the riverbanks and all the animal communities that rely on the river for water and food supplies.

Submerging millions of acres

Above the dams, large areas are often submerged (known as *riparian zones*). Large dams routinely submerge millions of acres, depending on the depth of the water. This displaces not only animals, but human communities can be uprooted as well.

Social conflicts are inherent to dam building. The products of the dam, namely electrical power, are generally enjoyed long distances from where the dams are located. Local populations become displaced, and they get very

moving current and generate electricity. Small-scale impoundment systems can also be built, although the economics are difficult to justify, and the regulatory burdens are becoming increasingly burdensome.

Most small-scale hydropower generators are used with batteries that store extra generated energy, when the dwelling's consumption does not meet the system production. And just like solar, one can tie into the grid with a small-scale system, thereby negating the need for batteries and also ensuring that if the hydropower system isn't functioning, for whatever reason, uninterrupted power will still be available.

The good and bad of small-scale systems

Small-scale hydropower has a lot of benefits if you're close to a suitable water resource:

- ✔ You can generate more kwH's per cost than any other alternative energy resource, in particular solar PV panels.

- ✔ No batteries are required (although they do make the system work better). You can assume your hydro generator output will be pretty constant, at least from hour to hour. Over the course of a year, you may have major variations.

- ✔ You can install a system of virtually any size power output if your water source is big enough, which is usually a requirement for economically viable applications. If your water source is small, power outputs will fluctuate quite a bit and may be zero for extended periods.

- ✔ You can generate power day or night, in any weather (freezing may cause problems, but you can usually design around them).

- ✔ Hydro systems have very long lives, are relatively trouble free, and require little maintenance.

- ✔ A submersible hydropower generator (moving water) for $1,200 gets you 2.4 kWh daily with a 9 mph stream. This is good economics, compared to other alternative electrical generating resources.

Small-scale hydropower also has drawbacks to consider:

- ✔ Complex electrical system designs and mounting schemes are difficult with water pressures pushing all the time. This is not a trivial system design to tackle, although do-it-yourselfers can safely do it if they're patient and willing to try things, then adjust, then try, and then adjust until they've finalized the best arrangement.

- ✔ Waterways can dry up in droughts and investments sit idle.

✔ Upfront costs are high, particularly for stationary water systems. In addition, not many hydro generators are sold because so few people have access to a good water supply.

✔ In addition to the hydro generator, you need a good inverter to convert the raw voltages from the alternator into the standard household voltages that will run your equipment.

Choosing to use small-scale hydropower

If you're interested in moving forward, you need to take some measurements based on your water source:

✔ **For moving water:** Measure the water flow and speed. *Flow* is how much water passes a given point in a minute. Generally, you won't be able to use the entire flow from a creek or river because that disrupts all wildlife functions. You can measure speed by simply tossing a stick out into the water and timing its travel over a given distance.

Measure the distance by pacing it out: one pace is around a yard. An approximate measure is good enough, because speed changes so much that an accurate measurement gives no better results.

✔ **For stationary water:** You need to devise a pipe system that produces maximum pressure. This is a complicated (albeit interesting) subject, and you'll need to find some detailed resources. There are two types: high fall (head) – low volume and high volume – low-head. The latter is similar in physics to an impoundment dam, where the turbine is located a distance below the level of water in the reservoir, while the former is similar in nature to the traditional, running water mill type of dam.

Just because you're next to a river or lake doesn't mean you have the right to use the water for generating power. You also need water rights (legal classifications), which can sometimes be confusing and contentious. Consult your county and state water regulating agencies for more details. You may need to consult with several different agencies simultaneously, which makes the permit process very burdensome, in some cases.

Oceanic Energy Sources

The oceans of the world provide two different types of energy: mechanical energy from tides and waves, and thermal energy from solar absorption. Thermal energy is relatively constant, while tides and waves vary quite a bit. The world's oceans cover around 70 percent of the planet, and so there is a tremendous amount of energy to be harvested, and it's renewable to boot. The technologies to harness this energy are still immature, but with such promise, a lot of investment money is being spent.

In the late 1970s and early 1980s, as a response to the Arab oil embargoes, a lot of research and development went into devising schemes to capture wave and tidal energy. In the U.S. alone, over $30 million was spent developing working systems. However, the till dried up in the 1990s and today the industry is small and mostly experimental.

With the price of oil rising, and concerns about greenhouse gases taking center stage, the amount of interest in ocean energy is on the rise once again. Recent studies indicate that wave and tidal electrical generating systems can be economically competitive with fossil fuel energy sources. If cap and trade policies are pursued, the economics of ocean energy become very favorable.

Tidal power generators

Twice every day the tides rise and fall, and there is a tremendous amount of energy to be tapped from the process. In fact, the process has been exploited for thousands of years, as waterwheels were mounted at the mouths of broad inlets and when the tides rose and fell the currents drove the wheels.

More recent embodiments include a sluice gate for capturing water at high tide within inlets. As the tide falls, water pressure build ups and a turbine/generator combination is mounted in the moving water to generate electrical energy. The system has the advantage of evening out the erratic flow of the tides so that power outputs can be obtained all day long, instead of only when the tides are shifting. This system is referred to as an "ebb" system because it works when tide is ebbing (going out to sea).

A tidal barrage reservoir is built like a small dam, with control gates that can be opened or closed, thereby enabling water to flow between two bodies at different depths. As tide comes in, the reservoir is allowed to fill through either an external channel, or through the turbine channel. When the tide is at peak, the level of water in the basin is the same as the ocean level. When tides fall, the control gates are opened and water flows through the turbine and the generator creates electric power. Some systems use dual turbines so that electricity can be produced both on rising and falling tides.

Tidal currents also flow in the vicinity of irregular shorelines, and these may also be tapped for energy sources. Rip currents flow as eddies near shorelines, and the water flows can be very powerful and consistent over the course of a day. Littoral currents flow parallel to shorelines in response to the rising and falling of oceanic tides, and depend on the shoreline composition. These can also be very powerful.

Following are some advantages of tidal-electric power:

- ✔ Tides are renewable, sustainable, and predictable. Unlike waves, tides won't disappear for days at a time.

- ✔ Some regions of shoreline feature very large differentials between high tide and low. These are ideal locations for tidal-electric generators.

- ✔ Tidal power generators produce no air pollution.

- ✔ A tidal barrage can serve a dual purpose: as a power generator and as a roadway over an inlet.

- ✔ Tidal barrage generators are easy to maintain. Existing units have lasted over 30 years.

- ✔ The turbines are located entirely beneath the surface, so they don't blight a beautiful environment.

Disadvantages of these systems include the following:

- ✔ Capital equipment is expensive. Up-front costs are very high.

- ✔ The technology is immature since so few of these have been built. This entails investment risk.

- ✔ Turbines can be difficult to install, since the best settings feature the harshest tides. Setting foundations is particularly problematic.

- ✔ Tidal power plants can affect the surrounding ecosystem. Marine life can be killed in the turbines. When large marine life get lodged in the turbine it can be a very costly problem to fix (not to mention an ugly problem).

- ✔ Failures of the system can result in flooding of the region around the basin.

Wave power generators

There are three main branches of research into wave power generators:

- ✔ Floats and bobbing devices are used to capture the energy in rising and falling waves. Existing small-scale units are capable of powering buoys and are reliable and consistent.

- ✔ Oscillating water columns in a cylindrical shaft increase and decrease the air pressure in the shaft as waves pass by. The pressure differentials are used to power a turbine, which is connected to an electrical generator.

- ✔ A wave-focusing scheme constructed near a shoreline directs waves into an elevated reservoir. When the water flows back toward the ocean once again, the pressure is used to spin a turbine connected to a generator. These devices have proven unreliable, and have largely been abandoned.

The most promising candidate is the oscillating water column scheme.

If you've ever been in an amusement park that features a pool with waves, you have the basic idea, because this process simply works in reverse. As waves pass by the cylinder, the water level rises and falls accordingly. When the wave is at a minimum, the pressure in the cylinder decreases and air is pulled from the outside ambient, through the turbine, which spins as a result. When the wave is at its peak, the reverse occurs. Air pressure increases and the turbine rotates in an opposite sense.

The power outputs of the turbine (and generator) vary quite a bit, but that's not a problem with modern high-powered semiconductor circuits. Grid ready electrical power is produced at a high efficiency.

Advantages of wave power:

- The turbulence of the oceans is a renewable energy source. In some parts of the world, waves are literally constant, and very powerful.

- There are no greenhouse gas emissions, nor any air pollution. Nor is there appreciable impact on the surrounding ecosystem. In the grand context of wave energy, these devices take very little of the ocean's energy. Their existence in an ecosystem is negligible.

- Wave generators are not expensive to install or maintain, although it is a tricky problem to get them anchored to the sea floor adequately.

- Wave farms can use combined outputs of individual wave generators to create large amounts of useable power.

- Wave generators have very low profiles (in contrast to off-shore wind generating systems which are visible for 50 miles in every direction).

Drawbacks of wave generator systems:

- When there are no waves, there is no electricity.

- They make noise, a strange sucking noise, when the air pressures change and the turbine spins. This may or may not be a problem, as some people like strange sucking noises.

- Big storms can be powerful enough to destroy a system. Hundred year systems are designed to withstand storms that only occur every hundred years, but hundred year storms do happen.

- Because wave generators feature such low profiles, boats may run into them inadvertently. This can be remedied by mounting a mast and flag overhead, but then the advantage of being relatively unseen disappears.

Chapter 11

Blowing Away with Wind Power

In This Chapter
▶ Understanding the fundamentals of wind power design and utility
▶ Getting familiar with large- and small-scale systems

Global air currents originate when the sun heats and cools various parts of the world at different rates. Winds result as nature attempts to even out the disparity. This is more than just an interesting tidbit. It also has important implications for wind as an alternative energy source.

It only takes a mild breeze to power an entire city with wind energy, so the potential to harness wind power is immense. Wind energy is clean, renewable, and homegrown (domestically produced). Above all, wind power is cost effective and among the alternative energy sources could see the largest growth due to its wide range of favorable factors and very short list of disadvantages.

In this chapter I describe the various technologies being used to harness wind power. And I analyze the pros and cons.

The Back Story of Wind Power

Humans first used wind power to move their boats well over 5,000 years ago. In the Western world, coarse windmills entered the scene at around 200 bc; these were used to process grains and other foods. The Chinese used windmills for the same purpose over 2,000 years ago. The Dutch refined windmill technology, which was used for draining swamps and grinding food. The English also built thousands of windmills for similar purposes.

As the technologies were refined, the number of applications soared. Windmills were used for irrigation, milling and processing of manufactured goods like spices, and woodworking products.

In the United States, heavy, cumbersome wooden blades were replaced by fabric sails (very much like the sails used on tall ships). Over 6 million windmills were built in the United States over the course of the next century to pump water on small ranches, mostly in western states where access to rivers and streams is limited. Nearly every western movie shows a scene with the windmill next to the house, and in fact the push west was made possible by these windmills because there was no other source of water, in most cases.

Electricity was first generated by wind power in 1890. This was made possible by larger, metallic blades that captured more of the potential wind energy and converted it into useable power. The first such windmill had a blade diameter of around 50 feet and output only 12 kW's of power, but it was the beginning or a new era in energy production.

As the federal government pursued large hydropower programs in the 1940s (refer to Chapter 10), windmills fell out of favor. The national electrical grid was being built and people didn't need localized power sources any longer. All they had to do was plug into a wall socket and they had their power. Fossil fueled generators were also being built on large scales, and the economics was much better than that for wind power. Plus, environmental concerns were minor back in those days.

As a result of the first Arab Oil Embargo in the 1970s, wind came back into favor as a locally grown, diversified energy source. Large-scale wind farms on an experimental basis originated in California in the 1980s as a result of federal and state tax subsidies. Since then, the technologies have advanced as computer aided designs allowed for much higher conversion efficiencies (the conversion between wind potential energy and output electrical energy). Power semiconductors also increased the electrical efficiencies of wind farms to the point where the technology could compete price-wise with fossil fuels.

Today, wind offers perhaps the greatest promise of any alternative energy source. The power is clean and efficient. There are no greenhouse gas pollutants and wind is available literally everywhere, at some time or another. And the cost is low.

Growing Markets for Wind Energy

Historically, wind power has only been able to fill very small market niches. The windmills that pumped water on western ranches fulfilled a vital function, one that could not be filled any other way, but there has never been large-scale water pumping using wind energy. And windmills are very localized in that the power they produce is generally available only at the powered shaft. Windmills used for grinding grain only worked on site. There was no way to store the energy, and there was no way to transport it (unlike fossil fuels which are easy to transport).

As electrical energy became feasible through advancements in turbine and generator technology, however, the scope of the markets has expanded considerably. Now wind power is not only a viable option for individual use, but it is a real option for large-scale energy production.

The world's circulating wind currents are virtually inexhaustible, and wind is available nearly everywhere in the world. Of all the alternative energy sources available, wind promises the most potential to most parts of the world. It's the most accessible energy source, even more accessible than fossil fuels (which must be extracted, refined, and transported).

Wind markets at a glance

Currently, there are four major wind energy markets, ranging from small-scale wind turbines that virtually anybody can install in their backyard, to large-scale wind farms capable of powering an entire city.

- **Small-scale production for remote locations:** In some locations, grid power (that is, power supplied by a utility company) is simply not possible due to long transmission line lengths. In these situations, wind power is an ideal solution because it's reliable and consistent, and the capital equipment costs are generally low. When wind energy is available, power is produced; when wind energy is not available, battery backup systems provide needed power.

  You can find remote wind energy used in off-grid homes, by telecommunications industries to power remotely located cell towers and switching systems, and in remote villages in undeveloped countries.

- **Hybrid systems:** Hybrid systems combine wind power generators with solar systems or hydropower systems to create a more diversified energy portfolio. Batteries are usually used as well. By diversifying the energy sources, a more consistent supply of power is ensured. These types of systems work in remote applications, but are also used with grid-connected systems as well.

- **Grid-connected systems:** Grid-connected systems are similar to off-grid systems — they're essentially small-scale systems designed to provide power for individual needs — except that the output power is connected to the utility grid. When the wind power system outputs more energy than the host building uses, excess power is available and can be sold back to the utility company, a situation that has resulted in the increasing number of small-scale wind power systems in use. (You can read more about the importance of net metering on the economics of wind power in the following section.)

✔ **Large-scale wind power systems:** A large number of individual turbines are connected to form utility scale power outputs that can power an entire city. You can find out more about large-scale systems in the section "Wind Farms: Utility Scale Electrical Production."

Economics of wind power

Since 1982, the cost of wind power has dropped by an order of magnitude, from around 40 cents per kWh to less than 4 cents now. This is less expensive than most fossil fuel energy sources. As a result, the global wind energy producing industry has doubled in size every three years since 1990. The rate of growth is accelerating. Expect to see a lot of big wind turbines in America's energy future.

Incentives for small-scale wind power systems

Small-scale wind power has proliferated due to net metering laws, as well as tax rebates and credits.

Without net metering, also called *intertie,* wind power economics would be hopeless under most circumstances. *Net metering*, selling excess energy back to the utility, is great for wind power economics because every wind system has a certain maximum capacity determined by the total amount of wind that blows past a turbine in a given day. While maximum capacity depends on a number of things — on a still day, system capacity is much less than on a windy day, for example — consumption is governed simply by what you need and want. In a typical household, power consumption peaks in the morning and evening while wind production peaks in the afternoon when the wind blows hardest. With a wind power system tied in to the grid, you can use as much power as you need, and your wind system merely contributes whenever it can. What you don't use from your wind system is all sold back to the utility.

The government has mandated that utilities not only allow wind power customers to connect to the grid, but that they pay the same amount for power they receive from wind systems as for the power they dispense to their customers. Without the mandates, the utilities would probably not allow wind intertie systems at all.

Investment tax credits have also been expanded to include wind production systems, so the use of wind turbines by individual homeowners will increase over the next few years. While these systems don't make much of a dent in the big picture, like solar, every little bit counts and adds up.

Declining costs of large-scale wind power production

The economics of large-scale wind power production are competitive with fossil fuel energy production, with costs per kWh as low as 4 or 5 cents. As technologies mature, the price of large-scale wind power has declined and wind now competes, without subsidies, with fossil fuel power production. In the last two decades alone, the cost of wind power has dropped around 90 percent. The average cost of utility wind power capital expenditures in 2000 was $790 per kilowatt (compare this to over $10,000 per kilowatt for solar power). When environmental costs are factored into the equation, wind is the cheapest form of alternative energy available.

Enhancing the economics is that wind power can be installed in small increments. Individual turbines may be installed one at a time, and then connected together when the time comes. This makes for easier capital equipment budgeting. Very few other alternatives can be built incrementally, with the exception of solar, but even then it's best to install a complete solar system upfront. Wind farms can begin producing energy when only a few turbines are completed, so the payback is very quickly realized compared to other investments (nuclear power plants take ten years to build, and are very expensive to boot).

Tracking wind power around the world

In 2003, there were more than 65,000 wind turbines cranking out over 39,000 megawatts of power the world over. These provided power to around 45 million people. This number is up over 300 percent from only a decade earlier. Government studies conclude that wind will grow by a factor of 15 in the next two decades alone, producing enough power to provide 6 percent of global electrical energy needs.

In Europe

Europe leads the world in wind production capacity, with around three quarters of all production. Europe also supplies around 90 percent of the world's wind generating equipment, and so their economy gets a big boost from the import dollars flowing in, not to mention all the high-tech, high-paying jobs that the industry supports. Currently, wind power comprises around 4 percent of all European electrical energy needs. Objectives call for this percentage to increase to 22 percent within 20 years. Germany is the global leader in wind power, with 14,600 megawatts of capacity, followed by Spain and Denmark.

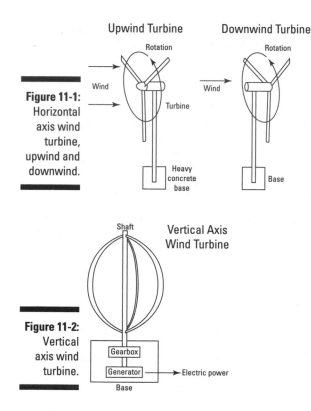

Figure 11-1:
Horizontal
axis wind
turbine,
upwind and
downwind.

Figure 11-2:
Vertical
axis wind
turbine.

Cut-in and cut-out limits

All wind turbines have cut-in and cut-out limits. *Cut in* refers to the minimum amount of wind necessary to enable the turbine to output power; there needs to be a certain amount to overcome the friction of the turbine. *Cut out* means the maximum amount of wind that is safe for operation.

Below cut-in and the turbine produces no electrical power. Above cut-out and the turbine is stopped and faced out of phase with the wind so as to reduce the possibility of damage. Optimal speed is usually just shy of the cut-out speed, so turbines are generally spinning very fast. The gearbox and controller ensures this optimal speed.

Stabilizing the mount and cooling

Heavy duty-scale wind turbines are mounted on tall towers, anywhere between 160 feet and 280 feet high. They are typically fixed into huge masses of concrete embedded deep into the ground. The torques on the towers can be immense, and some early experiments ended in disaster. Many towers are *guyed* (further stabilized using heavy steel cables attached into the ground around the turbine tower).

The blade bearings, electrical generator, and cooling means are mounted in a housing referred to as a nacelle. Cooling means are necessary due to the high power outputs generated in such a small volume. Inefficiencies create heat, which must be removed.

Maintaining rotational velocity

Typical systems are designed to rotate at a constant angular velocity of around 20 revolutions per minute (RPM) for the entire workable range of wind speeds. A constant rotational velocity is desired in order to maintain optimum aerodynamic performance.

As the wind velocity varies, the power output varies but the rotational velocity remains the same. This is accomplished by using a gear box (much like the transmission on a car). As a result, the electrical output of the generator operates at a constant 60 cycles per second, which is the U.S. electrical power standard. In European countries, the outputs are at 50 Hz, or cycles per second.

Accommodating wind direction

In order to work optimally, the turbine must be pointed directly into or away from the wind. The blades and nacelle rotate to accommodate changes in wind direction. Wind speed and direction are determined by an *anemometer* (a device that measures wind speed and direction), which is similar to the device that meteorologists use to predict weather patterns. Motors and gears rotate the nacelle to ensure it is facing into or away from the wind. Wind directions vary on a micro level, and the processor determines the most optimal *azimuth* (the orientation with respect to north/south — by definition, 0 degrees is due north, and 180 degrees is due south) at any given time.

In particularly strong winds, the blades are locked into place and the nacelle/blade assembly is rotated 90 degrees out of phase with the wind in order to minimize the torque forces that result from the wind.

Most heavy duty wind turbines operate with the blades facing into the wind (refer to Figure 11-1). This is called an upwind design. Other systems are leeward, or downwind, designs. There is ongoing debate about which type produces more output electrical power over the course of a windy day, and so far there is no consensus.

Picking the right spot

Location is critical, and there are areas of the country where winds blow nearly nonstop, 24/7. In order to optimize cost effectiveness, a wind turbine needs to be in the right location. The potential power output over the course of a day is directly related to the amount of wind velocity at a certain site.

Sites only a hundred yards apart can see big differences in performance, so there is a lot of engineering work done in site surveys. The potential energy increases by a factor of eight when the wind speed doubles, so it's highly desirable to find regions where wind is not only consistent, but very fast. In general, wind speeds increase as altitude increases, so it's desirable to position a turbine at as high an altitude as possible. It's also common to see wind farms mounted on the crest of hills.

Wind speed is defined in terms of meters per second, which is around 2.4 miles per hour. So a wind speed of 10 meters per second is equivalent to 24 miles per hour. Most large wind turbines are designed to operate at wind speeds from 3 to 4 meters per second up to 20 to 27 meters per second. Below 3 or 4 meters per second and the turbine doesn't spin. Above the maximum wind speed the turbine begins to lose efficiency, and output energy suffers. At too high a speed, the turbine can become damaged, and this is very dangerous in light of the size of modern turbines. Damage usually occurs at around 50 meters per second, which is a wind that makes it almost impossible to walk.

A large, modern wind turbine can output up to several megawatts at an optimum wind speed of around 25 meters per second. This is a considerable amount of power, given the size and footprint of a wind turbine site. Efficiencies are now on the order of 25 to 45 percent, with improvements in turbine design constantly pushing this number higher.

The best sites for wind farms are generally removed from population centers (people don't like a lot of wind, so cities generally aren't built in extremely windy areas). This means that most wind farms are connected to the grid via long transmission lines. Wind turbines are ideal offshore, where winds are consistent and strong. This also removes the turbines from sight and from human interference. The interaction with ecosystems is also minimized with off-shore sites.

The pitch of the blades

Wind turbine blade pitches may be varied so that efficiencies are enhanced (airplane propellers and helicopter blades also do this). When winds vary, optimal conditions for extracting the kinetic energy also vary. Intelligent sensors constantly adjust the pitch of the blades (the angle at which they intersect the wind), and when conditions are dangerous they completely offset the blades, tucking them away so as to minimize friction forces.

The total production of a turbine is also related to the surface area of its blades, and the torque produced by the blades, so the lengths and breadths are increased as much as possible. The longest blades being built today are over 300 feet in length. With the latest technologies, individual turbine outputs now exceed over 4 megawatts (MW). With towers over 30 stories high, and blades 300 feet long, a single turbine located in a windy location offshore can provide 1,400 homes with power. That's a lot of power and the scope illustrates the great potential of wind power, for there are literally tens of thousands of suitable offshore sites on American coastlines.

Other things to know about wind farms

Wind energy has a lot to recommend it: It's renewable, and the supply is virtually infinite. And wind farms offer several advantages as well:

- ✔ Turbines are maintenance free for long lifetimes. Very little labor is required to maintain a large-scale wind farm.

- ✔ Wind energy is distributed, in that it can be created all over instead of in a centralized location. The economics don't suffer when turbines are built far apart (aside from the wiring runs). It also increases the diversity of a nation's energy supply.

Environmental impact

They produce no greenhouse gases or other noxious pollutants, and there's no need for fuel transport into a wind farm, or waste transport out of the farm, but that doesn't mean that large-scale wind power has no environmental impact.

Specifically, large-scale wind turbines kill birds, bats, and insects. (*Note:* Small turbines used in residences and remote locations spin much faster than large turbines, but the surface areas are smaller, so the probability of a bird hitting it are also smaller.) A study done at a large-scale wind farm in Spain concluded that over 6,000 birds per year are killed by the turbine blades. As a consequence, a $1.6 billion dollar wind farm project in the U.S. was delayed by environmentalists (thereby proving that you simply can't make environmentalists happy, sometimes). In England, a government study resulted in the shelving of a large wind farm because it may cause the extinction of a particular species of coastline waterfowl. Outside of San Francisco, California, a study concluded that a large wind farm is responsible for the deaths of over 5,500 birds annually. This site is also a key site for nesting raptors, and the baby birds are being killed at a much higher rate than the adults. However, as the turbine blades become much larger, due to engineering advancements in materials and aerodynamics, the blades turn slower while producing the same amount of power. This will mitigate this problem.

However, to put this into perspective, consider that the Exxon *Valdez* oil disaster in Alaska was responsible for the deaths of over 1,000 times more birds than the California wind farm kills in a single year. Which means it would take nearly a thousand years for the wind farm to do as much damage

Chapter 12

Digging into Geothermal

Geothermal energy is the only completely renewable energy source from the earth itself. It's used to heat and cool buildings, as well as generate modest amounts of electrical power. A well designed geothermal system is perhaps the most environmentally friendly source of energy possible. And perhaps best of all, geothermal energy is completely domestic in supply, reliable, renewable, sustainable, and versatile in that it can be used to both heat and cool directly as well as generate electrical power.

More than 70,000 times more energy is available in the earth's upper crust than all of the potential energy in the fossil fuel reserves previously discovered. Even accessing a small fraction of what's available would make a major dent in our reliance on foreign oil. And with each new geothermal source brought on line, our pollution problems are mitigated. While geothermal will never be a major player in the grand energy drama, every little bit counts, and geothermal is best when it's done small scale, so it's available to a lot of users.

In this chapter I give you a lowdown on geothermal energy sources and how they're used to generate useable energy, both big and small-scale.

Geothermal Energy Basics

"Geo" is a Greek word meaning "earth," and "therme" means "heat." Geothermal energy originates in the inner mantle of the earth as hot molten magma (liquid rock) circulates upward while surface groundwater seeps downward. The magma heats the water and forces it back up through cracks and faults. In liquid form, this geothermal energy is referred to as *hydrothermal*. If you were to drill a well down into this heat, you'd tap into an incredible amount of free, completely natural energy.

The earth's core is estimated to be around 8,000 degrees Fahrenheit (nobody's actually measured it, of course), and this heat is constantly radiated out toward the earth's surface. In fact, the heat bubbles right up to the surface in some places (hot springs and geysers, for instance). Turns out Mother Nature is truly a hot babe, when you get right down to it (or right down into it).

For every mile of depth below the surface, the temperature increases around 80 degrees Fahrenheit. In some areas, the temperature rises much faster, making available a lot of heat energy just below the surface. Tapping into this energy requires drilling down into the earth's crust, much like the drilling processes that are used to extract oil and natural gas. In some cases, the hydrofluids are under high pressures, and there is no need to pump the fluids to the surface because they will just spew out. In other cases, active pumping systems must be used to extract the fluids up to the surface.

Where you can find geothermal energy

Geothermal energy sources are available literally everywhere in the world. Given the role of magma in generating geothermal heat, you probably won't be surprised that volcanic regions offer the best potential for geothermal energy. Geologically stable regions, or high altitudes, often offer poor potential as geothermal energy sources. Still, some high-altitude regions that don't have any volcanic activity — like Wyoming and South Dakota — have very pronounced geothermal potentials.

Classifications

Geothermal fields are classified depending on the temperatures, which vary between 194 degrees Fahrenheit to over 300 degrees Fahrenheit:

- **High-grade sources:** Above 400 degrees Fahrenheit. Temperatures can go as high as 1,300 degrees Fahrenheit, which implies a tremendous amount of energy recovery. High grade sources are most abundant in the U.S. western states and Hawaii.

- **Medium-grade sources:** Between 300 and 400 degrees Fahrenheit and generally found in the southwestern U.S.

- **Low-grade sources:** Between 212 and 300 degrees Fahrenheit and may be found anywhere.

Different temperatures require that different engineering methods be used to exploit the energy. As temperatures increase, pressures increase as do danger levels from the hydrofluids. To prevent scalding and the unintentional releases of superhot steam, the piping systems need more integrity and more safety interlocks.

Tapping into hard to reach places

While geothermal energy sources are available everywhere in the world, some areas — those deep beneath rock terrain, for instance — are simply too difficult and expensive to tap into.

As technology progresses (in particular, the same drilling technology constantly being advanced by fossil fuel recovery) geothermal will become more widespread and economical.

Dealing with water sediment

Water is an ideal medium for moving thermal energy. Yet the water itself can present a problem. Because the water has been trapped inside the earth's crust for so long, it generally has a lot of sediment and mineral buildup, which causes engineering difficulties that are hard to overcome. Because of the high sedimentary buildup, geothermal fluids are referred to as *brines,* and they're very reactive: They tend to corrode and decay metallic based equipment very quickly. Although ways have been devised to remove the impurities from the fluids, they are expensive and time consuming.

Brine contains several potential sources of environmentally damaging contaminants: Hydrogen sulfides (H_2S), ammonia, methane, carbon dioxide (not that again!), sulfur, vanadium, arsenic, mercury, and nickel. *Closed loop systems* (those that return the pumped fluids back into the ground) return these materials to the earth, where they came from, while *open loop systems* (those that dump the used hydrofluids into a river or onto the ground) expose these contaminants to air and surface water.

All these contaminants come from the earth; the problem with returning them to the earth is their relative concentrations. Ecosystems are all local, and when geothermal sources are tapped, it's important to return the pumped sources of heat energy close to the well sites. For this reason, closed systems are favored by environmentalists.

Geothermal uses

There are two main ways to use geothermal energy: to produce electricity and to directly heat (or cool) individual homes and businesses.

✔ **To produce electricity:** Geothermal energy sources can be used to produce electricity through a variety of methods: dry steam production, flash steam production, and binary-cycle production. You can read more about this process in the section "Creating Electrical Power With Geothermal Sources."

✔ **To heat and cool individual homes and businesses:** Directly heating individual structures with geothermal energy is a very simple process: The hydrothermal liquid is simply piped through a radiator system within the building.

Although direct use requires a completely different process than electricity production, both still use the same fundamental physics of heat transfer. Heat is moved via conduction, which means that the heat moves from the pumped fluids into radiators and coils that then transfer the heat into the air within a dwelling.

Economics of geothermal energy

Geothermal energy has a lot going for it in terms of economics: There are no raw fuel costs, nor are there any transportation difficulties. Maintenance costs are generally very low, depending on how much sediment is present in the heated liquids drawn from the earth. A typical geothermal electrical generating plant is very self-sufficient, compared to other options. Once it's up and running, a geothermal plant can output at full capacity consistently for up to 20 years.

The main factors affecting the economics of a geothermal energy source include the following:

✔ **Flow rate and pressure:** How quickly and with how much pressure the hydrofluid moves impacts the type of piping system that must be used.

✔ **Ease of getting to the source:** How easy or difficult it is to access the heat, given the geological features — rock, soil, and so on — in the area.

✔ **Location:** Some locations are much more difficult to tap into. For instance, if there is a lot of rock, the drilling process is very costly.

✔ **Demand level:** How competitive geothermal is, price-wise, compared to the market price of other energy options.

✔ **Capital equipment costs:** The costs associated with drilling machinery and pumping equipment. Drilling through rock requires much more expensive equipment. Drilling long distances also requires more expensive equipment.

Small plant capital costs are estimated at around $2,500 per kW of output, depending on the accessibility of the geothermal source. Large plant costs are much lower, at around $1,600 per kW; a 30 megawatt plant, for example, would cost around $40 to $80 million dollars to build. Costs per kWh of production vary anywhere between 8 cents to over a dollar (this latter number is very high).

In individual homes or businesses, using geothermal energy directly to heat or cool can reduce an HVAC (heating, ventilation, and air-conditioning) bill by up to 65 percent, although the upfront equipment costs for residential heat pumps are generally higher than for other types of equipment.

Creating Electrical Power with Geothermal Sources

Some power plants uses geothermal sources to generate electricity. At present, only around a quarter of a percent (0.25) of the world's electrical energy comes from geothermal sources. Table 12-1 lists the top five nations' electric geothermal outputs.

Table 12-1	Top Five Nations for Geothermal Output
Country	*Output (in megawatts)*
U.S.	2,200 Megawatts
Philippines	1,900
Indonesia	800
Italy	780
Japan	570

In the U.S., geothermal provides 2,200 megawatts of electric power, and comprises 10 percent of nonhydropower, renewable electricity production. This is enough to power 1.7 million homes, or the equivalent of four nuclear power plants. However, only four states use this power: California, Hawaii, Utah, and Nevada. California alone derives 6 percent of its power from geothermal sources.

The actual production of geothermal is a very small percentage of its potential, however. Estimates by the Department of Energy (DOE) indicate that up to 10 percent of the U.S. electrical power could be generated by geothermal, with minimal environmental impact. The reason geothermal is still a small player is that upfront capital equipment costs are high compared to alternatives, particularly fossil fuel.

Alaska and Hawaii hold the most potential for geothermal exploitation. Most of the western half of the United States offers economical potential for geothermal power plants. A study done by the United States Geological Survey concluded that over 20,000 megawatts of power are available in the western states, and this number could expand by a factor of five if engineering advances were made that could make use of liquid temperatures lower than the current requirements.

Types of electricity-generating systems

There are three major ways to use geothermal energy to produce electricity on a large scale (small-scale generators also work, but aren't economical compared to other alternative energy options, so they are far and few between). The type of system used in a particular application depends on the quality of the source, chiefly the temperature and quantity of the heated fluid.

Dry steam power plants

Dry steam reservoirs are huge manifolds of very hot steam within the earth (in the absence of water), generally at low pressures, although sometimes the pressures can be extremely high. While these conditions are rare, it's very easy to use this energy source to extract useable power. If possible, these are the most economical geothermal sources. Capital equipment costs are lowest for this type of plant because the process for extracting the heat is so direct and straightforward.

A production well must be sunk deep enough to tap into subterranean rocks with temperatures higher than the boiling point of water (212 degrees Fahrenheit). A pipe system channels the dry steam to the surface, where it's fed into a turbine/generator combination which produces electrical power. The steam, which has cooled considerably as it passes through the turbine, is injected back into the earth's core via a separate well called the injection well. *Note:* The production well and the injection well need to be far enough apart so that they don't interfere with each other (if they are too close, the fluid will simply circle around).

Flash steam power plants

Flash steam power plants tap into liquid heat reserves in the earth's crust. The water pressures are very high, so it's easy to draw the liquid to the surface. The temperatures of useable wells are generally above 360 degrees Fahrenheit, which makes for some dangerous conditions.

As with dry steam (explained in the preceding section), a production well is used. This time, instead of drawing dry steam, it draws heated liquid. Around 40 percent of this hot water immediately converts into steam when the pressure is released in the flash tank, and this is used to drive the turbine/generator, which produces electrical power.

These systems require more capital equipment and backup equipment than the dry-steam plants because the machinery is more complex (basically it's two separate systems interacting with each other), and as a result don't produce as much useable power per dollar of investment. Maintenance is more expensive as well.

Flash steam power plants, when placed on a coastline, can be used to desalinate water supplies for drinking and irrigation. This is a completely natural result of the fact that distillation occurs when water is boiled to vapor.

Binary-cycle power plants

If a liquid geothermal source temperature is below around 360 degrees Fahrenheit, a binary-cycle system is in order. These systems have two independent closed loops; the production well/injection well loop, and the generator loop. A heat exchanger transfers the heat from the well loop into the generator loop, where the turbine/generator produces electrical power. Having two separate systems generally implies inefficiency.

Binary-cycle systems are economical when there is enough hydrofluid to justify the cost of the capital equipment. Sometimes wells dry up, and this results in a big waste. For this reason, these types of generators are more risky than the other two types.

Things to know about geothermal power plants

Geothermal energy has quite a bit going for it: The supply of energy is virtually limitless. There are no toxins or pollutants generated in the process of either using geothermal energy directly or as a way to generate electricity. No fuel has to be transported to or waste material transported from the site — an incredible advantage over other large-scale electrical energy sources.

Geothermal energy is considered renewable since the source is so immense. But, strictly speaking, when the heat is removed and the liquid or steam is pumped back down into the earth, there is a permanent change in the earth, and so geothermal is not renewable. But the change is so minute in the grand scheme of things that it's virtually negligible.

Finding a suitable site

As stated earlier, the supply is practically limitless . . . if a suitable site is identified and tapped. Finding suitable sites is sometimes very difficult. And if the suitable site just happens to be near an urban area, it's hard to get the necessary permits.

Locating suitable sites is more of an art than a science, in a lot of cases. It's not much different than locating oil or natural gas reserves, and there's no way of measuring with any accuracy the true potential. So investments inevitably take some guesswork, and this increases the financial risk.

Self-sustaining plants

Geothermal power plants are self-sustaining: They don't need external sources of energy for pumping or generation. A small amount of electricity may be required to get the initial pump phase in gear, but once the system is producing its own electricity, it's completely self-sustaining.

In addition, the operating costs are minimal, aside from routine equipment maintenance. Power plants can be run with almost no labor costs.

Plant safety

Plants are safer, in relation to most fossil fuel power plant generators. There is almost no risk of explosion and leaks aren't nearly as noxious as with fossil fuel generating plants because the leaked material comes from the earth and goes right back to the earth.

However, on rare occasions flammable vapors and liquids may come out through the production well, and these can be very dangerous and difficult to extract out of the heat source. When this happens, extra costs are incurred to ensure safety.

The risk? Running cold

There is an inherent risk in building an expensive geothermal plant because they sometimes "run cold," which means that the heat source begins to disappear. Natural changes in the subterranean environment occur regularly (as the earth shifts). These can manifest in two ways: The temperature of the geothermal source can begin to fall, thereby making useable energy production more inefficient, or the quantity of fluid available can decrease. Pressure may also decrease, meaning it takes more energy to extract the heated fluids.

Using Geothermal Energy Directly: Heat Pumps

There has been a shift in the use of geothermal from producing electrical power to using the heat directly to warm homes and buildings. Using geothermal energy directly is very simple: You just use the hot water to heat something, either a home or the ground. There's no processing required and no electrical production. Today, direct use systems provide over 18,000 megawatts of power in over 60 different countries. The current worldwide direct geothermal capacity heats over 3 million homes.

Direct-use geothermal systems can be used in many ways, including:

- Heat a single home, or an entire community.

- Warm the water used on fish farms, thereby increasing productivity rates. Warmer water also allows for the propagation of species that would normally not be found in certain regions. (Hmm . . . As an environmentally conscious citizen I wonder if this is really a good thing? Perhaps this is fooling with Mother Nature a little too much?)

- Heat greenhouses the world over. This may be one of the most common uses of direct source geothermal because greenhouses take a lot of heat due to their poor insulation properties. Other greenhouse heating options are prohibitively expensive, in comparison.

- Pasteurize milk.

- Dehydrate fruits, vegetables, and grains.

- In agricultural applications, geothermal heat is used to warm underlying soil, making for better crop production. The heat also serves to sterilize the soil, to some extent. In some cases, both irrigation and heating can be accomplished in a single operation.

- The health of livestock has been improved by using geothermal heat to both sanitize and temperature regulate holding pens.

- Heat can be pumped through sub-surface piping systems beneath roads and walkways to prevent ice-buildup in the winter.

Direct geothermal heat, when engineered properly, may be used in several phases, increasing the overall efficiency of the process to impressive levels. For instance, when first extracted from the ground, the highest temperature fluids can be used to drive power plants. The exhaust fluids, which are now cooler in temperature, can then be used to heat buildings and finally the even more cooled fluid can be pumped into the ground beneath agricultural fields.

It may be true that of all the energy sources available, direct use geothermal energy is the most efficient and cleanest source of all. The issue is one of availability: either you got it or you don't.

What direct-source systems need

When geothermal heat sources are available, they can be used to channel hot water directly through buildings to provide heat in a completely natural, inexpensive fashion. These sources are generally found at around 50 degrees

Although air-source heat pumps require an expensive upfront investment in equipment, they're the less expensive type of heat pump to install. And because installation is relatively easy, you can get a wider variety of qualified installers to give you a bid. Competition is always a good thing.

Ground-source heat pumps

Ground-source heat pumps are very similar to air-source heat pumps, except they use the earth as their heat exchange medium. They are more efficient than air-source pumps because the earth is much more consistent in terms of temperature. *Note: Geo-exchange* is the common name for ground-source heat pumps. It's just a matter of terminology, but you'll see both terms (*geo-exchange* and *ground-source* heat pumps) in use.

Geothermal ground-based systems circulate a water-based solution through a loop system that's buried underground. Water (lakes, streams, and so forth) can also be used as the exchange medium for ground-based geothermal systems, but this is rare because water temperature changes more than land temps, and water is harder to access than earth (very few people have a ready access to a suitable supply). However, when a good source of water is available, it can make the process very efficient.

With a ground-source heat pump, fluid is pumped from the ground (or from a suitable water reservoir) through the compressor and the heated fluid passes through a coil arrangement inside the dwelling. A fan draws air through the intake vent (which normally includes filters of some kind) and over the coil, thereby heating the air. The fluid, which is now cooler, is pumped back into the earth via a discharge well.

Heat pumps in extreme climates

In extreme climates, a ground-sourced heat pump with a radiant heating system in the floor is a great combination because you can simply heat water directly, and this heated water is then pumped through the floor grid of the radiant system. You can also use that hot water generated by the ground-source heat pump in your domestic supply, thereby killing two birds with one stone (although killing birds is probably not strictly in sync with the green mantra). Radiant heat is very quiet, and in combination with the quiet ground-source heat pump, your home will be virtually silent and without moving air. (*Note:* If you use a forced-air duct system in conjunction with a ground-source heat pump, you need extra equipment — which costs more — to convert the hot water into hot air, as well as to move the hot air through the ducts.) Be aware, however, that using radiant heat in your floor doesn't work very well for cooling in the summer.

Ground-source heat pumps work very well and can be installed in almost all climates because ground temperatures are very stable compared to air temperatures, so the heat source is very consistent. But the capability comes at a price: Ground-source heat pumps are expensive because they require extensive piping systems be dug into the ground. These systems allow maximum contact with the earth that's used for the heat exchange. The most common ground-source heat pump options are open- and closed-loop piping systems.

An open-loop system uses existing ground water, while a closed-loop system has its own dedicated fluids and the heat is moved via exchange. In many areas, the ground water is very sedimentary (because of mud, minerals, salt, and other reasons) and therefore unsuitable for open-loop systems, which are less expensive to install.

The efficiency of heat pumps

The efficiency and power output of a heat pump is a function of the outdoor temperature (the ground or air used as the source). When outdoor air temperatures are extremely low, heating is less efficient, and in very cold climates air-source heat pumps may simply not be able to output enough heat to warm a dwelling. Likewise, when outdoor temperatures are extremely hot, the cooling efficiency of a heat pump suffers. With ground-based heat pumps, the efficiencies are relatively immune from the weather because ground temperatures do not vary much over the course of a year. (The effects of the seasons are nonexistent about 6 feet underground. In northern states the ground stays at a consistent temperature of around 50°F, which is a lot warmer than the air aboveground in the winter months. In hotter climates, the ground can be as warm as 70°F year-round.)

In general, heat pumps do not heat and cool nearly as quickly as separate units, in particular combustion heat sources can provide a lot of heat very quickly. However, due to their inherent efficiencies, heat pumps that run constantly are often more economical than combustion heaters which constantly cycle on and off.

In addition to heating and cooling, heat pumps can also provide:

- Domestic hot water supply
- Air filtration
- Humidity control

The ability to provide all of these functions simultaneously makes heat pumps very economical and efficient in many applications.

The commonly used rating for a heat pump is the "ton," which has nothing to do with weight. This term is a holdover from the days when refrigeration units were used to make ice. A 1 ton unit could make 1 ton of ice per day. To bring this into modern terminology, a 1 ton heat pump can generate around 12,000 Btus of cooling per hour at an outdoor temperature of around 95 degrees Fahrenheit, or 12,000 Btus of heat at 47 degrees Fahrenheit outdoor temperature.

Advantages of heat pumps

Heat pumps have their share of pros and cons. On the plus side, heat pumps tout the following characteristics.

- ✔ **They offer steady, even heat and cooling.** Fluctuations in temperature are smaller compared to traditional combustion and air-conditioning equipment, and this increases comfort. With combustion sources, the walls and floors and furniture are often cold while the air is hot, and the air tends to move faster with combustion type heaters because of the increased power outputs. Moving air tends to make people feel colder than the actual temperature (convective cooling removes heat from the skin).

- ✔ **They take up less space than traditional combustion equipment** because of their two-for-one advantage (they can both heat and cool). Plus, there's no need for a chimney or venting system because there's no combustion (they still may use ductwork to distribute the heat, but radiant systems are more common because they are a natural partner).

- ✔ **They're safer and cleaner than other options.** They don't create on-site air pollutants like smoke, carbon monoxide, and so forth. Nor do they create ashes or creosote. No combustion occurs, and none of the components become extremely hot. A malfunctioning heat pump won't endanger a building's occupants; it'll just freeze them out.

- ✔ **They require more maintenance than some other options.** Systems require a number of parts, and heat pumps are more complex than combustion systems. If you look inside an air-source heat pump, you'll see what looks like the space shuttle. Nevertheless, heat pump technology is mature and getting better all the time. If you buy new equipment, the quality will be good in terms of both design and implementation.

- ✔ **They require electricity — sometimes a lot.** This not only affects energy efficiency, depending on how you're getting your electricity (nuclear source, combustion source, and so on) but it also means that in a power outage you're outta luck.

Chapter 13

Exploring Biomass

*I*n many parts of the world, biomass is the chief source of energy. A third of the world's population burns forest products, animal dung, and other organic matter as part of their daily ritual of living. In the industrialized world, biomass plays a large role in home heating and it is seeing increased use of biomass for electrical generation and as a substitute for conventional fossil fuels.

The growth of the biomass industry is perhaps the highest of any of the alternative energy industries due to the versatility of applications and the universal availability of biomass raw products which would otherwise be wasted. For instance, wood pellets are made of waste products from the forest industry, and methane gases are produced from cow manure.

Common sources of biomass include trees and grasses, crops and crop residues, aquatic and marine plants, manure and sewage, and landfill (paper, cardboard, cut brush, and discarded food). With advances in technology, there are even more possibilities on the horizon.

This chapter looks at both the current state of biomass energy and its future as a significant player in the alternative energy field, and explains large- and small-scale biomass energy options, as well as biofuels.

Biomass Basics

Traditionally, biomass has been used for heating and cooking. But this traditional use of biomass products only taps the surface of the potential. Machines are being developed that can burn municipal garbage (certainly an inexpensive energy source) and agricultural waste like worn tires, plastics, sewer byproducts and forest and corn husks and the like. Efficient machines are also being developed that can create large-scale electrical production from biomass products.

Biomass is a renewable form of energy because it derives from the photosynthesis process in which plants convert the sun's radiant energy into carbohydrates (as opposed to hydrocarbons, which comprise fossil fuels). When plants are grown specifically for use as biomass, they also constitute a form of energy storage. Biomass can be described as "stored solar energy."

In general, biomass refers to organic matter of any kind, whether forest products or agricultural production or simply the plants that grow on the surface of the earth. These are broadly referred to as *carbohydrates.* Almost any organic-based materials can be utilized in biomass energy schemes. These include wood products from trees, crop residues (material left over after agricultural production for food sources), animal wastes (poop), aquatic plants like seaweed, landfill gas (from rotting garbage), and municipal and industrial wastes (like sewage).

Note: In the next chapter I describe wood burning. While wood is biomass, under the strict definition of the word, it merits its own chapter because it's so widespread and easily accessible.

Biomass uses

The uses of biomass vary. On a small scale, residential homes burn biomass like wood pellets and corn in specially designed stoves. On a large scale electrical power plants burn methane gas produced by compost piles of biomass materials. Methane can be produced on large scales, and fed into the pipelines that feed urban areas with natural gas supplies. Ethanol is a biofuel derived from corn, and it's increasingly being used in transportation around the world. Here's a more detailed breakdown of how common biomass energy sources are used:

✔ Wood and grasses are directly combusted to provide heat for boilers which can drive turbines and produce electricity. Steam locomotives originally burned wood before that fuel source was supplanted by coal. The most widely used source of wood biomass for combustion purposes is the waste from the timber and forestry industry, in paper mills and lumber processing plants. The waste is formed into pellets that are burned in residences and businesses for heat. Willow trees, switch grass, and elephant grass are also grown specifically as biomass fuels for this purpose.

✔ Corn, poop, and wood pellets are burned in residential stoves to provide heat. Other sources of biomass may be used as well. Some stoves can burn pretty much anything that's thrown into them, although efficiencies vary quite a bit. And some fuels create quite a stink in the process (poop, for instance, stinks on both ends of the process).

✔ Crops and crop residues (particularly corn) are used to produce ethanol, a liquid which is commonly added to gasoline. Other grains such as wheat, rye, and rice are used to produce biofuels. Soybeans, peanuts, and sunflowers are used to make biodiesel fuel. Biofuels can be used for electric power generation as well as transport.

✔ Microalgae, found in lakes, is fermented into ethanol or composted to produce methane gas.

✔ Animal waste from farms and ranches, as well as human waste, is composted to produce biogases (double yuck, but it works and makes a lot of sense, along with a lot of stink).

✔ Landfills are harvested for paper, cardboard, and other useful organic products which can be burned to create electricity.

✔ Old tires can be burned to produce both heat and electrical power.

Some biomass materials are more efficient than others in creating useable energy. Energy density is important — wood and corn are energy intensive: They hold a lot of potential energy in a unit's worth of weight. But even biomass products that aren't energy intensive can be useful: After all, they're already here but aren't being utilized in any way. Poop, for instance, is a biomass product in great abundance in livestock production, but for the most part it's not being used for anything other than fertilizer. The vast majority of poop is left to rot where it is "deposited" (how's that for a euphemism?).

Converting waste to energy

Converting waste into useable energy is one of the most promising areas for biomass because the environmental impact is so small. Waste for energy conversion includes all carbon-based products generated through normal human activities (garbage, for instance, contains a lot of potential energy). U.S. agriculture and industry produce over 280 tons of carbohydrate wastes each year. Imagine if all of this waste could be efficiently converted into energy.

There is a downside, of course. Most waste products are difficult to combust without some processing and can result in a range of pollutants that may be even worse than those emitted during the fossil fuel combustion cycle. Agricultural waste, on the other hand, holds a tremendous potential for producing large amounts of energy without damaging the environment in the process. Corn, sugar cane, and rice offer the most promise. The byproducts (husks, shavings, and so on) of producing these foods are relatively straightforward to harvest with little extra investment in time and money. Traditionally, farmers have had to pay to get these waste byproducts hauled off to a landfill, but with advances in biomass machinery farmers can now charge for these materials. This has considerable economic benefits for entire communities.

Growing energy: Fuel crops

In addition to using the waste of agricultural crops, it's also possible to grow crops specifically for energy. The type of crop determines the type of use:

- **Starchy plants,** such as corn and wheat, are good for liquid petroleum supplants such as ethanol and biodiesel fuels. (These crops are also good for human consumption because they have relatively simple organic structures that are easy for the human digestive system to break down.)

- **Cellulose-based plants** (such as trees and grass) are best for heat and electrical production via direct combustion.

Historically, energy crops have been largely ignored in relation to food crops, but interest is increasing as the economics of energy grow more expensive and complex.

When crops are grown solely for energy production, species that offer the best efficiencies and least pollution potentials are selected. Energy crops can provide up to six times more energy potential than the energy required to produce them. With energy crops, the ratio of useable-output-energy to wasted-energy-to-produce-the-useable-energy is very large compared to many other energy producing sources.

Some types of trees, such as poplars, maples, sycamores, and willows, reflexively grow back when they're harvested, so the process is naturally sustainable, with minimal labor and expense. It takes a relatively short time — around three to seven years — for these trees to grow big enough to be reharvested. In the future, there will be huge farms dedicated to growing and harvesting these types of trees, and converting the raw materials into useable forms.

Energy crops are also environmentally friendly. Many of the plants used to produce methane gas are advantageous to the environment. Switch grass, for instance, reduces erosion and provides a natural habitat for wildlife. They grow deep roots and actually add nutrients to the soil. Current research funds are being directed toward developing energy biomass plants that blend in with the natural ecosystem, and therefore disrupt the environment minimally.

Compost piles: Biomass at work

A compost pile is an example of biomass at work. You throw garbage and any old organics you have lying around into a special hopper, and the decomposition process creates heat by breaking down the materials in an anaerobic process. You can use the final product any number of ways, including fertilizing your landscaping and burning in a stove. For that matter, burning autumn leaves is a biomass combustion process.

The benefits of biomass

Biomass is renewable and sustainable, if grown under the right conditions. There are a number of advantages with biomass.

- ✔ Global warming is caused by the buildup of carbon dioxide in the atmosphere. Biomass combustion, while it does release carbon dioxide, does not add to the buildup. When fossil fuels are combusted, carbon that has been locked up inside the earth for millions of years is released into the environment. When biomass is combusted, the CO_2 emitted is taken up by the plants or trees that grow the next crop of biomass fuel, effectively creating a closed loop cycle for the carbon.

- ✔ Biomass supports local agriculture and forestry industries, thus reducing our need for foreign oil.

- ✔ Biomass supports the development of new domestic industries (jobs) such as biorefineries that produce fuels, chemicals, and other bio-based products.

- ✔ It's versatile: Biomass production can be either local, in one's backyard, or it can occur on a large scale. Big farms produce corn for use in ethanol, and the refining and processing and distribution industries are tremendous. On the other hand, anyone can grow some biomass in their backyard and use it for various purposes such as heating and composting, or even eating.

- ✔ Biomass left to rot on the ground releases just as much carbon dioxide into the atmosphere during the decay phase as biomass that is combusted, so it's effectively carbon neutral.

Of course, all is not rosy: Critics contend that there will be major environmental ramifications when huge areas of land are dedicated to growing specific, sometimes non-native, varieties of plants on a large, constantly evolving scale. This is hardly letting nature take its own course. There are instances where certain species of plants have become so invasive to a new environment that they completely take over the region's ecosystem, rarely for the better.

Looking at the future of biomass

Currently, biomass is used to heat homes and buildings, as well as to generate electricity. Biomass can also be used to make liquid biofuels for transportation and a range of other useful chemicals and bioplastics, as later sections in this chapter explain.

Developing biomass industries should be a national priority, and movements in that direction are starting to materialize. In the state of Maine, for example, the Fractionation Development Center is promoting a "forest biorefinery." A study concluded that within four years, Maine could produce half of the transport and heating energy needs that the state currently consumes. Plans call for some 60 biorefineries, employing 7,000 people. The energy dollars stay completely local, and the jobs would help a weak economy reliant on foreign oil.

The U.S. Department of Energy (DOE) and the U.S. Department of Agriculture also have plans to develop biomass production. Plans envision a 30 percent offset of current fossil fuel energy sources with biofuels by 2030. In this scheme, biomass energy will replace 5 percent of the nation's electrical production, 20 percent of transportation fuels, and 25 percent of all chemical production. This will require over a billion tons of dry biomass feedstock annually, which is a fivefold increase over current levels.

The U.S. currently has the resources in forest lands and agricultural lands to support this kind of growth in the biomass industry. And the program prescribes sustainable growth methods that will constantly replenish the raw sources of biomass without disrupting the production of food supplies. Roughly 75 percent of the raw biomass needed to support this plan would come from croplands, and 25 percent from forests.

Despite these developments, biomass will never be able to completely supplant the use of fossil fuels. A 2003 study concluded that it would take 22 percent of all the plant matter that grows on the surface of the earth to sustain the world's current energy consumption through biomass energy producing systems. This is roughly twice the world's current agricultural production level. Nevertheless, biomass can make an appreciable dent, since the benefits are so widely spread across the economy.

The Industrial Revolution was characterized by a consistent movement away from biomass carbohydrates toward fossil fuel (hydrocarbon) energy sources. The current trend toward biomass represents an unprecedented reversal of this process. Perhaps it's not true, as the statistics in Chapter 3 so compellingly suggest, that humankind is inalterably destined to use more and more fossil fuels.

Large-scale Biomass Power

Plant and animal wastes can be processed to produce methane, which is a key component of natural gas (the fossil fuel variety). While methane releases harmful greenhouse gases into the atmosphere, when it's produced using biomass, the carbon is reabsorbed by trees and plants as they grow, so that the net release of greenhouse gases is zero with biomass methane.

The process of turning biomass into power

The methane produced through biomass can be used in any number of applications in the exact same way that conventional, fossil fuel methane is used. Figure 13-1 shows the methane production cycle from biomass products.

To produce energy with biomass, a huge compost pile is left to rot. The copious amount of methane gas that is produced from the rotting material is channeled into the combustion chamber in a combined cycle methane power plant. The compressor pressurizes incoming air flow, which combines with the methane to produce a controlled combustion that powers the turbine. The turbine is connected to an electrical generator. The heat from the combustion is further directed into a boiler where steam is produced. The steam pressure turns another turbine/generator combination and another source of electricity is produced. The two electrical sources are combined into one, and this is fed into the power grid.

Generating electricity this way is very efficient and environmentally sound since the net release of greenhouse gases is zero. The economics are favorable, particularly in light of the fact that the biomass products used in the compost are very low in cost, and locally produced. Biomass generators can be located near agricultural and livestock production farms, so transport of the raw compost materials is minimal.

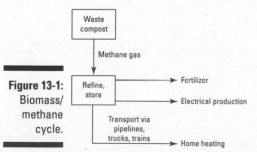

Biomass / Methane Cycle

Figure 13-1: Biomass/ methane cycle.

The joy of cooking: Making your own methane

You can convert cow manure into a very energy-dense gas by building a special fermentation processor. The malodorous raw material is pumped into a *digester silo* where it is heated to 95°F, at which point fermentation is activated. Anaerobic microorganisms break down the organic matter to produce a gas that's two-thirds methane and one-third carbon dioxide. The odor is similar to rotten eggs. The gas inflates a large plastic storage bellows. Methane is drawn off into a combustion-powered generator to produce electricity, which can be used to power nearly anything. All you need is a big pile of cow poop, but that's not hard if you have a bunch of cows. In fact, it's unavoidable.

Advantages and disadvantages of biomass power plants

In addition to the general advantages of biomass, biomass power plants offer these benefits over fossil fuel combustion plants:

- The biomass used in the plant is very inexpensive, compared to other energy sources.

- Biomass is low in sulfur, so there is very little SO_x emission when it is burned properly. Modern power plants burn the fuel very efficiently, and this in turn controls sulfuric emissions.

- Biomass power plants can operate continuously, unlike solar and wind farms which rely on the vagaries of the weather.

- Methane can be produced on small sites, and then transported to a central location. Even small farms can produce quantities of methane gas, and this is becoming more common throughout the country as farmers and livestock growers are learning that they can recoup some of the costs of their operations by producing and selling methane. They can also use the methane in their own operations, which often results in impressive economic gains. Small- and large-scale methane production diversifies the nation's energy sources.

Of course, there are also a few problems associated with biomass combustion plants, the chief one being ensuring that waste products are disposed of properly, which can sometimes be very expensive. If processed properly, biomass waste products can be used as fertilizers and clean landfill materials.

Another significant issue is ensuring that the biomass is combusted properly: If not, it can release even more harmful products into the environment than fossil fuel combustion does. A study concluded that a municipal incinerator in New Jersey released 17 tons of mercury, 5 tons of lead, 750 tons of hydrogen chloride, over 80 tons of sulfuric acid, 17 tons of fluorides, 580 pounds of cadmium, and over 200 different types of lethal dioxins into the air each year. The remaining ash is also toxic and has caused landfill problems. Burning waste needs to be done very carefully in order to extract real benefits.

Another problem, particularly for those who live near or work in these plants is the foul smell. Workers must have an inordinate tolerance of "rotten egg" smells, or worse.

Biomass Fuels

Liquid fuels are produced from biomass like corn and wheat. Ethanol is the most common form of fuel, followed by biodiesel. I describe these two fuel types in this section. There are other possibilities, but their usage is so rare that I won't get into the details.

Ethanol

Ethanol is basically liquid oil produced from biomass products. It's chiefly used as an additive to liquid fossil fuel products like gasoline, although there are internal combustion engines that can burn pure ethanol. Ethanol mixed with gasoline is called *gasohol* (an ugly word, but there it is). Conventional autos and trucks can use up to a 25 percent mixture of ethanol without loss of performance.

Ethanol has historically been more expensive than gasoline, and so its use has been limited. But with gas prices rising, while ethanol production costs fall, the economics are becoming more favorable. Government subsidies of the ethanol industry have also improved the economics and it's now common for most of the fuel that consumers pump into their autos to contain some ethanol. Around 30 percent of the gasoline sold in 2007 in the U.S. contained ethanol. Ten percent is a typical mix proportion.

Brazil is the global leader in ethanol production, at around 50 percent of the world total. Sixty percent of their vehicles are powered by a 22 percent mixture of ethanol/gasoline, and the remainder of their vehicles use pure ethanol, which is widely available. The U.S. produces a quarter of the world's ethanol, and this percentage is rising due to government subsidies of the industry. In Europe, government mandates require that all vehicles use a 6 percent mixture by 2010, which represents a ten-fold increase in demand. This will further exacerbate the world's food shortage.

According to government studies, ethanol could provide 10 percent of the nation's transport fuel needs by 2010, and up to 50 percent by the year 2050. These are big numbers, and the impact on both pollution and reliance on foreign oil will be impressive if the goals are met.

Ethanol's benefits

Only 35 percent more energy is available from ethanol as was required to produce it, so the conversion efficiency is not impressive. And ethanol is much less energy dense than gasoline, so it takes more weight and volume of ethanol to achieve the same mileage. But while ethanol is less energy dense than gasoline, when it's added to gasoline it improves engine performance in two ways:

- ✔ Ethanol contains a lot of oxygen (in comparison to gasoline) and this improves the combustion in an engine. It also increases efficiency and reduces emissions in the process.

- ✔ Ethanol octane levels are higher than those of gasoline, so engines perform with less knocking. This makes refueling less expensive because with gasohol there is no need to buy the higher grade fuels (high octane fuels, normally called premium).

Beyond improved engine performance, ethanol offers other significant political and environmental benefits that make the drawbacks well worth the price. It's renewable, sustainable, and environmentally neutral. It relieves the burden of foreign oil and is produced locally so that jobs and money stay in a community. Expansion of the ethanol industry has resulted in many new American jobs.

Ethanol's problems

Most ethanol is produced from corn kernels because corn is in such abundance. In 2006, 12 percent of all corn production in the U.S. went toward making ethanol. This has resulted in an increase in food prices the world over, and there is a lot of political debate about whether this effect should completely stop corn/ethanol production. Developing countries are the hardest hit when food prices rise, so it may be said that the net effect of ethanol production is starvation in developing nations.

Another downside is that corn production requires a lot of energy, water, and labor, plus modern machine intensive farming practices degrade the environment in many ways. Environmentalists don't generally support the production of corn based ethanol, instead preferring other biomass bases like sugarcane and switch grass.

Making ethanol

Just in case you're curious, there are three basic ways to make ethanol:

✔ Pyrolisis is a thermochemical process that condenses carbon compounds through high temperature heating. This is the oldest and most widely used technology but it's also the most energy intensive, so the efficiency of producing ethanol in this way is poor. Furthermore, the process needs to be controlled very carefully, which requires a lot of oversight and expensive capital equipment. Other products of pyrolisis are oil, acids, water, and various gases, all of which may be used for other applications.

✔ Fermentation is a biochemical process that uses microorganisms like yeast to effect an anaerobic conversion of sugars into alcohol, and is very similar to the process used to make beer (too bad you can't use beer

in your gas tank because it's cheaper than gas nowadays, plus it has a very nice dual purpose). In this process, sugar cane works better than corn, but sugar cane supplies are nowhere near as robust as those of corn so this is relatively rare. Fermentation plants are inexpensive to build, but the overall conversion efficiency is not very good. The amount of energy required to produce ethanol through fermentation is not much more than the energy that can be extracted.

✔ Synthesis converts biomass into a gas, using a liquefier arrangement. This gas is mostly used in power plants in the same way that methane is used. This method holds the most long term promise, and research and development funds are being directed toward developing better processes and machinery.

Finally, in the U.S., ethanol is produced in only a few states located in the Midwestern corn belt. But the population centers are on the coasts, and so transportation issues are burdensome. Fossil fuel refineries are generally located nearer to the urban centers so as to minimize transportation costs, but there's no way that can be accomplished with ethanol.

Biodiesel

Biodiesel is an organic liquid that may be used as a substitute for fossil fuel based diesel fuel. It's typically made of soybeans, which are a huge crop in the U.S. Seeds from many different types of plants are ground down and processed to make biodiesel. Vegetable oils like used kitchen grease may also be burned directly in diesel engines, with only minor processing and refinement.

Biodiesel is currently twice as expensive as fossil fuel, so its use is not as widespread as that of ethanol. But if the cap and trade policies of reducing carbon emissions are implemented, biodiesel will become cost competitive.

Biodiesel may be mixed with conventional diesel fuel in any proportion without adverse engine performance. Biofuels are less explosive than diesel, so they are also safer.

Small-scale Biomass for the Masses

As mentioned previously, not all biomass energy production is large-scale. In fact, traditionally, biomass has been used on a small scale by individual homeowners.

You can install a biomass stove in your home, and take advantage of local energy sources while helping to preserve the environment. Stove technologies have improved dramatically over the last decade. Here's what you need to know about burning biomass:

✔ Biomass can be a very clean-burning fuel or a very dirty-burning fuel. Knowing what you're doing is the key to success. Do it right, and you can burn biomass more cleanly than you can burn fossil fuels. Do it wrong, and burning biomass can create a lot more pollution and ash than most other options. Burned right, biomass is very efficient in terms of producing heat from the potential energy in the fuel.

✔ Many biomass products are commonly transported long distances because the packaging makes transport and storage easy. Wood pellets, for example, come in compact bags that can be stacked on pallets and moved easily. Also, because the energy density is high with wood pellets, you get more bang for the buck.

Getting the lowdown on biomass stoves

Biomass stoves and furnaces are designed to burn biomass fuels. In appearance, biomass stoves are similar to wood stoves, and they transfer heat the same way. But biomass stoves are set up to burn at different temperatures and may require more or less oxygen in order to achieve the highest efficiencies.

Biomass stoves are generally more efficient than wood-burning stoves because biomass combustion processes are well-controlled. These stoves have automatic augers that feed the fuel into the burn chamber at a controlled rate. You simply load a hopper, and the fuel feeds down into the fire only as needed. You can control the feed rate by means of a thermostatically controlled switch, and you can mount the thermostat either by the stove or in a remote location.

You can heat your entire home with a biomass furnace installed in your basement. Or you can heat a single room. You can buy a very cheap biomass-burning stove and get inefficient performance, or you can spend more and get excellent convenience and efficiency. How you plan to use your stove determines the size and operation of your equipment. Well before you buy any stove, ask at the stove shop what your options are and then go home and consider the details, as well as how you'll implement them once you get going.

Choosing a fuel

The decision on which type of biomass to use is generally dependent upon which types of biomass are available in your area. The Midwest favorite seems to be corn stoves. In the Northwest, you find a lot of pellet stoves (pellets are made of byproducts of lumber processing, like bark, sawdust, twigs, and so forth).

Burning pellets

The main operational advantage of pellet stoves is that they're very easy to use. You fill the hopper with manufactured pellets and the rest of the process is largely automatic. A thermostat keeps the temperature at a preset point, and an auger feeds material into the combustion chamber so that you don't have to worry about it. You can fill the hopper once a day or even less often if it's not too cold out, and your home will stay at the set temperature. Pellet stoves can usually burn different types of materials as well, such as corn, coir fiber (from coconut husks), and nutshells. With very little work, you can get an even, consistent heat source.

Many types of pellets are made from byproducts of other lumber processing ventures, like sawdust and ground wood chips. Some pellets are even made from corn stalks (usually wasted since they have no nutritive value) and nut hulls, or from other crops like switch grass. Pellets are like recycled products, and even better for the environment because of it.

Two key advantages of burning pellets for heat are

✓ **Pellets are renewable and produce the lowest emissions of any solid fuel.** Wood pellets are neutral in terms of their effect on global warming because wood left to rot on the forest floor creates as much carbon dioxide as wood burned in a stove. And pellet combustion in a biomass stove is controlled much more accurately than wood combustion in a wood stove, so the combustion is more thorough and the efficiencies are much better. In addition, most of the industrial companies that manufacture pellets replenish the trees they cut down to make pellets.

✔ **If burned properly, pellets can be very clean and efficient.** Smoke is minimal, and the burn is clean enough with pellet stoves that you often don't need to vent the exhaust all the way up the chimney, so installation is easier and cheaper. There is also less creosote, which is safer for the home and cleaner for the environment.

Burning corn

Corn is the second most popular biomass fuel because it's so readily available in many parts of the country. To be effective in combustion, corn must be very dry (unlike the stuff you eat). Corn can also be very cheap and clean, if you have a ready supply of quality product.

Other grains also work in essentially the same fashion as corn. Wheat, barley, rye, sorghum, and soybeans can be dried and burned at low cost and with low environmental impact. As with corn, this is really a question of access. People in cities don't burn soybeans simply because they can't store soybeans and don't have a ready supply.

Like other types of fuel burned in biomass stoves, corn can be a very consistent heating source because the stoves use automatic feeds and large hoppers. It's also very clean burning and low on the pollution scale. Burning corn and other grains offers other advantages. First it's the cheapest renewable fuel (at least for the time being). However, with the price of ethanol (which is made of corn) increasing, the cost of raw corn also increases. And, because corn grows fast and furious, it's perhaps the best renewable energy source on the planet.

Burning corn creates some unique problems that non-corn burning homeowners don't face.

✔ You have to buy a stove entirely dedicated to burning corn. In some regions, these stoves are very rare, and servicing can be hard to find.

✔ Storing corn takes up more room than storing wood pellets because the energy content of corn isn't quite as good as that of wood pellets. In other words, you have to burn more corn to get the same amount of heat that you'd get from pellets.

✔ Corn has a limited lifetime, unlike other fuels. It may rot, or rats may eat it before you can burn it — a problem you have to address proactively or risk being overrun.

✔ Your home will have a sweet smell, not unlike popcorn. Depending on your tastes, this may or may not be a good thing, although people with corn stoves get used to the smell and seem to like it.

Chapter 14

Burning Wood

Firewood is the most plentiful source of biomass energy on the planet, and the most widely used because of it. This is not going to change. While burning wood the wrong way harms the environment and wastes money, burning wood the right way makes sense, and it can be very economical, especially if you have a ready supply of firewood on your property.

The one truly outstanding feature about wood fires is this: Wood left to rot on the forest floor creates as much carbon dioxide in the air as wood burned in a fire. And the carbon dioxide released from wood burning is a closed loop in that new plant growth uses that carbon dioxide to form new wood. With fossil fuels, the carbon comes from the ground, and has been sequestered there for millions of years. From a pollution standpoint, wood burning is environmentally neutral (although to be precise, wood burning does entail a lot of soot and other harmful gases). If global warming is of critical importance, wood burning does have its place in the grand scheme. And since it's so widely utilized, it fits into the alternative energy scenario on a scale that's readily accessible to most anybody.

There are no large-scale wood burning power generating plants; burning wood as an alternative to fossil fuels is a personal, mostly residential thing. For that reason, I get very practical in this chapter. I describe the big picture, but I also provide a lot of down-to-earth tips on how to burn wood the best way, to get the most energy possible and to harm the environment as little as possible.

What the Blazes! Cleaner and More Efficient Ways to Burn Wood

Humankind has been burning wood since time immemorial. In fact, it might be said that civilization began when people learned how to burn wood wherever and whenever they wanted. Fireplaces have been a traditional source of heat for thousands of years, and nothing is quite as cozy and inviting as a good open fire.

For a long time, wood-burning appliances (by that I mean stoves and fireplaces) comprised the vast majority of "hearth appliances" in a typical household. But that's changed dramatically as gas-fired stoves have grown in popularity. In 2000, over two thirds of all hearth appliances were gas stoves, and 2 percent were pellet (biomass). That still leaves a lot of woodstoves, and in some areas of the country, particularly rural where wood supplies are plentiful, wood burning still represents the vast majority of heating energy sources.

Increasing demand, improving economics

As with all other alternative energy sources, the Arab Oil Embargo of 1973 increased demand for woodstoves. A much larger surge in demand accompanied the Three Mile Island accident in 1979. Technologies improved, and wood burning efficiencies increased by a factor of two, so the economics improved as well. Since the amount of pollution produced by a wood burning appliance is directly related to how much that appliance is burning, air pollution also decreased.

Improving efficiencies

In the early 1980s a great deal of research was done to monitor the levels of pollution from wood burning activities of all kinds. The industry developed catalytic converters and better combustion chambers as a result. Pollution levels have declined considerably, and with a modern wood stove the amount of pollution released into the atmosphere is not much more than a good natural gas fired stove.

The U.S. Environmental Protection Agency banned all sales of high emission wood stoves beginning in 1992, but there are still a huge number of old stoves cranking away. They never die because they're made of heavy steel and are basically very simple, with no moving parts that tend to break.

Over time, customer demand for catalytic converter stoves decreased because of technical issues related to cleaning the catalytic converters on a periodic basis. As a result, most new stoves do not have catalytic converters, but instead use clean-burning techniques perfected in New Zealand. The clean-burning non-catalytic stove is the most widely sold in the world, presently.

Open Fireplaces: Cozy and Costly

Picture a fire crackling in an open hearth. Add the overstuffed chair, the steaming mug of cider (or goblet of wine). Throw in the golden retriever sleeping in the soft glow of yellow firelight, if you like. Makes you feel all warm and snuggly, doesn't it? Well, here's a blast of cold reality: Open fireplaces are about the least efficient and effective heating scheme you can imagine. Why? Because they draw cold air into a home to feed themselves and, due to the chimney effect (hot air rises aggressively), most of the hot air they generate goes right up the chimney.

Here's the typical process.

1. **You start the fire.** This entails kindling, matches, and some "blowing" to get things going. And maybe some swearing when things don't get going the way you had in mind.

2. **You open the damper.** If a cooking vent or bathroom vent is on in your home, or if the HVAC system is on, cold air may rush down the chimney, creating a back draft and causing smoke to seep out into your living space.

3. **The chimney effect kicks in, and then smoke begins to rise up the chimney, and the fire grows hotter and brighter.** Flickering orange flames, the norm, mean the wood isn't burning very efficiently. In fact, flames of any kind are inefficient because the wood is not burning very thoroughly.

4. **Because it needs fresh oxygen, the fire in your fireplace pulls up to 500 cubic feet per minute through windows and air leaks in your home.** If somebody opens a door to the outside, cold air will rush in because the pressure in your home is less than it is outdoors.

The end result is that your home may actually be getting colder, not hotter. There's no denying the romantic element. But at what price?

If you just can't live without the romantic open flames or you're stuck with what you've got, you can achieve better efficiency with these tips:

- ✔ When lighting your fireplace, open a door or window nearby so that the rush of air a fire takes before it settles down to a more static state will only move through a small portion of your home. Once the fire settles down, close the door or window.

- ✔ Burn hot, blazing fires because the combustion is more thorough. Don't burn little fires.

- ✔ Use andirons to lift the logs above the floor so that during the burn cycle the hot cinders fall down onto a bed of coals and burn better and more thoroughly.

- ✔ When you're not burning a fire, keep the damper closed. When the damper is open and a vent fan or the HVAC system in your home is turned on, air is drawn down the chimney. Not only does this result in cold air, but it could also mean a big stink and unhealthy air.

Wood-burning Stoves: An Efficient Alternative to Fireplaces

Stoves enable you to burn your wood more efficiently. They're designed such that the wood burns in a specially designed burn chamber, which enables you to achieve much more efficient combustion (by controlling the burn temperature) plus a more efficient dispersion of the heat into your living space.

Before deciding what type of stove you want to install in your home, check with your county building department for requirements, specifications, and so forth. These may make a big difference in your decision on which type you can buy. Also, check with your insurance company: They probably have some dictates you need to follow as well.

Comparing open- and closed-vent systems

There are two important types of venting systems in stoves: closed vents and open vents. In an open-vent system, air is drawn from the room and used for combustion, and then the exhaust is vented up the chimney. In a closed-vent system, air is drawn from the outside and vented back to the outside.

So which is better? Efficiency and safety-wise, a closed-vent system is:

✔ **An open-vent system is always less efficient than a closed-vent system.** The air that an open-vent system draws from the room inevitably comes from the great outdoors through cracks and leaks in your home's envelope (thereby making the home colder).

✔ **Closed-vent systems are safer than open ones.** Naturally, you want to seal your home to a high degree in order to make your heating efforts more efficient. However, in doing so, you may cause a carbon monoxide or carbon dioxide hazard with an open-vent system. Or you may cause oxygen depletion, because the fire is competing with the human inhabitants for oxygen.

Free-standing stoves and inserts

Aside from open- and closed-vent systems (see the preceding section), stove types feature another distinction: They can be either free-standing or inserts. *Free-standing stoves* simply sit in a corner of your room, without walls or bricks surrounding them. An *insert* fits into an existing open fireplace slot (a very good way to go from poor efficiency to good efficiency while still using the expensive decorative elements of your existing fireplace).

Free-standing stoves are inherently more efficient than inserts because, all else being equal, more of the generated heat makes it into the living space. Depending on the space they are mounted into, some inserts are completely surrounded by brick and mortar and do a very poor job of getting heat into the living space. Most of it goes up the chimney. (Even at that, an insert is still much more efficient than an open fireplace.) On the other hand, free-standing stoves take up more of your home's square footage, may be an eyesore, and are more likely to cause burns. In the summer, they're just plain superfluous sitting there looking black and bulky.

Types of wood stoves

In producing new clean-burning stoves, manufacturers have taken one of two routes: one that uses a catalytic combustor and one that causes more complete combustion.

While older wood stoves had efficiency ratings of 40 to 50 percent, today's certified stoves boast efficiencies of 70 to 80 percent. As the efficiency goes up, the pollution a stove creates generally goes down (for one thing, when efficiency is high you burn less fuel; for another, when you burn fuel much more completely, there are fewer pollution byproducts).

Stoves with catalytic converters

Some wood stoves incorporate a *catalytic converter*, which burns off smoke and fumes that would otherwise exhaust or drift up a chimney. The catalytic converter consists of a honeycomb-shaped substrate coated with a catalyst, usually a precious metal such as platinum or palladium. When the smoke passes through the honeycomb, the catalyst lowers the smoke's burning temperature, causing it to ignite. The result is a more efficient burn, less smoke up the chimney, and less pollution. Most stove manufacturers now offer catalytic models in which the combustor is an integral part of the stove. It is also possible to retrofit some older stoves with converters.

Starting a fire in a catalytic stove is more involved than doing so with a conventional stove. You have to open a bypass damper when you either start or reload a fire. The damper directs the combustion gases directly up the flue (bypassing the catalytic element) until the temperature is high enough (350°F to 600°F) to get the combustor kicked into gear. This process is referred to as *lighting off*. A catalyst temperature monitor is normally included to let you know when to move the damper.

Catalytic stoves require a lot more maintenance than their conventional counterparts. The catalytic convertor needs to be changed periodically. Otherwise, you could actually end up with a very inefficient fire. These types of stoves are more expensive, and probably not worth the cost unless you put a very high premium on pollution.

Modern, conventional stoves

While catalytic converters offer the best pollution performance, they have seen limited sales because of the problems inherent to cleaning and maintaining the catalytic converters. New stoves feature more controlled burn chambers that produce far less emissions than their heavy steel counterparts that are so common in older homes.

The oxygen intake to the combustion process is regulated so that the wood burns at a higher temperature, which results in more heat output with less pollution byproducts. The higher temperatures also result in more heat per unit weight of raw wood fuel. The units are heavier, and have more insulation around the combustion chambers. While the emissions are not quite as good as with a catalytic type stove, the performance is far better than with old stoves. The new style stoves are also capable of burning a wider variety of woods.

Important Points about Burning Wood Efficiently

Wood-burning stoves are the most common stove simply because wood is available virtually everywhere. Because the supply of wood is self-replenishing, it has the further advantage of being renewable. Wood-burning certainly has the most tradition on its side, so more equipment is available, as well as a wider range of raw wood resources to choose from.

Wood-burning stoves are banned in some areas. In other areas, burning is banned on particularly dirty air days. In these areas, utilities generally offer rebates and subsidies to convert to cleaner systems. Check with your local air resources board.

A wide variety of heating appliances are available at literally any price you can pay. You can find old stoves for next to nothing, and they work well enough although the performance may not be real good. Be sure to check the efficiency ratings of the stoves you're considering, because they may vary quite a bit. With some stoves, most of the heat is lost up the chimney. Other types of stoves are very efficient at channeling the heat into a room.

Enhancing efficiency by burning the right wood

Some types of woods are better than others for burning. Two factors determine the suitability of a given wood: whether it burns well and generates heat, and what kind of mess it makes in the process.

If pollution is your concern, insist on burning either dead trees or doomed trees (those that have been removed for building developments or other similar reasons). Don't use trees that have simply been cut down for burning. If you do, insist that a new tree be planted in its place.

Hardwoods versus softwoods

Hardwoods are almost always better than softwoods for burning because you get a lot more Btu per unit weight and volume, which means a lot less work on your part per Btu generated. Hardwoods also deposit less creosote in your chimney and vent system, making them safer and cleaner for the environment. Some people are suckers for the fact that softwoods are so much cheaper than hardwoods, but in fact, hardwoods are more cost effective. Check out www.hearth.com/fuelcalc/woodvalues.html for a calculator on different types of firewood and the Btu you can expect from them.

Seasoned wood

Any wood you burn must be properly *seasoned,* or dried out (wet wood smokes a lot, won't burn at a hot temperature, hisses and pops and may be dangerous, and puts a lot of crud into your vent system). To season wood, split logs as soon as possible into the size that will fit into your stove and stack them in a dry spot for 6 to 18 months. Pile the wood so that air can circulate. Hardwoods take longer to dry than softwoods do. Humidity and temperature also affect drying times.

Green wood is freshly cut from a live tree. It will smoke like crazy and stink and probably won't burn very thoroughly. Plus, a lot of water will steam out of it, resulting in sparks when you open the door to the burn chamber. How can you tell whether wood is green? See whether it's brittle — if so, it's not green.

Although you shouldn't burn green wood, you can buy green wood in late spring, when prices are very low because demand is down. It will dry over the hot summer months and you'll save money. In fact, if you can, buy a whole gob of green firewood at the same time and you'll get a great discount.

What you should never burn

Although burning hardwood is better than burning softwood, the latter isn't out of the question. Some things, on the other hand, should *never* be burned in a wood stove:

- ✔ **Garbage, plastic, foil, or any kind of chemically treated or painted wood:** They all produce noxious fumes, which are dangerous and polluting. If you have a catalytic stove, the residue from burning plastics may clog the catalytic combustor.

- ✔ **Trash (paper or plastic):** Paper wastes tend to make a very hot fire for a short period of time, encouraging the ignition of any creosote deposits in your stovepipe. And synthetic wastes, such as plastic wrappers, produce acids when they burn. Acids make for a very short stove lifetime.

- ✔ **Manufactured logs (like Duraflame):** This is extremely dangerous. They burn too hot and can cause a fire when you open the door.

Avoiding problems by burning the wood right

The potential for pollution is high if you don't use your wood stove properly (you don't maintain the proper burning temperatures, for example, or you don't keep the stove cleaned out). Burning wood properly (to get the most efficiency) requires you to pay attention to the process and continually manipulate the wood stack as well as the damping levels. Follow this advice for safety, efficiency, and convenience:

- In the firebox, avoid placing pieces of wood in parallel directions, where they may stack too closely. Always try to get some air movement between pieces of wood.

- Adjust the burning embers periodically with a poker to keep your fire burning properly. Unfortunately, there are no rigid guidelines; learning to use your stove optimally takes practice. Just play around with it.

- If you're looking for the most heat fast, vary the position of the wood to maximize the exposed surface area of each piece of wood. If you're after a long term burn (overnight, unattended) minimize the space between the pieces.

- Only use wood properly sized for your stove's fire chamber. Smaller pieces burn faster and hotter.

- Always keep the door closed when attending to the fire — when the door is open, too much oxygen enters the burn chamber.

A good technique for producing a relatively clean, long burn (overnight) is to load your stove with a mixture of partially seasoned and well-dried wood about half an hour before bedtime. Leave the damper open to give the fire a good start; then damp the strongly burning blaze down for the night.

Maintaining a wood stove

Wood stoves require chimney maintenance. Burning wood creates creosote gunk on your vent or chimney liner, and this can catch on fire if you're not diligent. Softwoods are much worse in this regard than hardwoods.

Maintaining a wood stove means having the chimney swept. This entails going up onto the roof (or wherever the vent terminates at the top) and forcing special brushes down that clean away the built up creosote and tar.

The consistency of the burn

In gas stoves, combustion can be tuned to a high degree, and the stove will always work the same way because it's a closed process that doesn't require intervention (plus the fuel is very consistent). On the opposite extreme are wood stoves, which vary in performance for a number of reasons:

- The wood composition and quality varies quite a bit. Even if you buy the same species, such as oak, and the wood is cut from trees in the same forest, the variations are tremendous.

- Wood is often piled high, and the stuff on the bottom gets muddy. This affects combustion quite a bit, not to mention the mess it makes.

- Stoves vary. The same wood burned in different stoves will combust differently.

- Over the course of the burn cycle, raw wood turns to ashes. Ashes build up and change the dynamics of the combustion process.

- Weather can affect how a stove burns by changing the way the chimney draws. If your home is oriented just right (or just wrong), you may experience severe back drafts on windy days.

Once the heat is generated in the stove, the next issue to contend with is how the stove gets the heat into the room.

A stove that is not damped excessively (and that has a well-designed chimney of factory-built, insulated pipe) may not need a sweep for an entire season. On the other hand, even a comparatively well-installed system could need cleaning as often as every two weeks. Your stove's instruction manual will provide explicit instructions on how to tell when cleaning is needed and the best way to go about it. After installing a new stove, be sure to check the stovepipe for creosoting every two weeks until you become accustomed to the heater's behavior. (Any deposits over 1/4-inch thick indicate that the pipe needs attention.)

You can monitor the accumulation of creosote and other deposits in your stovepipe by tapping on the sections with a metal object. Once you're used to the ringing sound that a clean pipe makes, the dull thud of a dirty one will be distinctive.

It's a good idea to have your stove system inspected by a pro at the beginning of each burn season.

To monitor the operation of your wood stove, check the exhaust coming out of the chimney. You should see the transparent white steam of evaporating water only — darker, opaque smoke will be just slightly visible. The darker the color of the exhaust, the less efficiently your stove is operating.

Getting the hot air where you want it

If all you want to heat is a single room, your options are simple because you don't need to move air in a complicated fashion and you can buy a small stove. If you're planning on heating a number of rooms or an entire home, you need to figure out how you're going to move the warm air from the stove to the adjoining rooms. When it comes to moving air, you have several options:

- ✔ **Ceiling fans:** These are a cost-effective way to reduce fuel bills and even out temperature variations. If your home has high ceilings or an open loft, most of the heat will rise to those areas and eventually migrate out through the roof. A well-placed fan will move this hot air back down into your living spaces.

- ✔ **Ecofans:** Ecofans offer a very efficient and effective way to move the hot air your stove is generating. An Ecofan (check out www.realgoods. com or enter "Ecofan" in your search engine) exploits the simple laws of thermodynamics to move a fan blade without the use of electricity or external power. You set the fan on top of your stove and after it gets hot, the blade spins and moves up to 150 cubic feet per minute (cfm) of air. Ecofans cost over $100, but over the long term they're a lot cheaper than using ceiling fans, and they can make your wood stove up to 30 percent more effective (I said effective, not efficient) by distributing the hot air.

- ✔ **Your HVAC fan system:** Most HVAC furnaces (the kind that use forced air) have a switch setting that allows the blower to run without having the heat on (Fan Only is a common label). If you run your HVAC system in this mode, it will distribute the heat generated by your stove through-out your entire home.

Stove Safety Guidelines

Stoves obviously pose a lot of dangers. With any type of stove, you can start a fire, create smoke hazards, or get burned badly. To avoid potential danger and the legal liabilities that may come with it, heed the advice in this section.

Ensure proper installation and venting

Wood-burning stoves can be very dangerous if they're not installed properly. A properly designed chimney (or vent system) is a prerequisite for any safe stove installation. The flue must be made of a suitable heat-resistant mate-rial and it must be separated from combustibles by a specified distance (see county codes and installation specifications for a particular stove).

Rules regarding masonry chimneys used to vent a stove state that they must be lined — usually with clay flue tiles, which should be mortared at the joints with refractory cement and separated from the stone, brick, or block work. The interior surface shouldn't show signs of chipping. Likewise, the exterior masonry and joints should be sound.

Don't install any type of stove yourself. A whole host of details go into the installation of a stove, and failure to pay attention to any one of them can lead to a safety problem.

Use of a wood stove may increase your fire insurance premiums. You must tell your homeowners insurance agent that you have a stove (they'll probably ask). In many cases, they'll come to your home to inspect the installation. If things aren't as they should be, your insurance company may refuse you or give you a set amount of time to get it up to code.

Bring an existing chimney up to date

The likelihood of an existing chimney meeting all of the criteria outlined in the preceding section isn't very good (if you're in doubt, have an expert come in and take a look). If the chimney is deficient because it's too large, lacks a liner, or is in poor repair, you have these options:

- ✔ **Bring it up to snuff (relining and so forth).** If you choose this option, let a pro do the job.

- ✔ **Abandon it.** For about the same cost as relining, you can install a factory-built, insulated metal chimney. This is a good idea because technology has improved quite a bit, and you'll be getting better performance along with better safety.

- ✔ **Install a new stovepipe in an existing chimney.** This solution not only eliminates the problem of the deficient chimney, but it also gives you the option to look into a closed system (a stove, in other words).

When using a fireplace chimney for stove exhaust, the entry to the fireplace must be sealed, or, if the connector joins the flue above the fireplace, the chimney must be plugged below the point of junction. This precaution not only prevents burning embers from cascading down into your fireplace (and onto your floor), but also maintains the proper draft.

Most county regulations prohibit the passage of stovepipe through any floor, ceiling, or fire wall without special feed-throughs. You may, however, pass your pipe through either a wall or a floor/ceiling if you use a factory-built insulated chimney. If you don't have a suitable masonry chimney, this expensive venting option is your only choice.

Follow manufacturer's instructions

Make sure you know and follow the safety and operation guidelines outlined in your stove's instruction manual. If you don't have an instruction manual, get one online or contact the stove's manufacturer.

✔ Contact your county building department to find out what the applicable codes and guidelines are in your area. If you don't follow code and you cause a fire or other danger, you could have an insurance claim denied or even find yourself in legal limbo if there's collateral damage.

✔ Call your homeowners insurance company and ask them what type of requirements they have for stoves. Many insurance companies will insist on coming out to your home so they can inspect your stove firsthand. (So many stoves have been installed inadequately that they can't just take your word for it.) Sometimes the insurance company's standards are stricter than the county's, and you may find your insurance canceled or your rates increased if you don't mind your p's and q's.

✔ Keep in mind that the fuel you burn in your stove is just as combustible while it's in storage as it is when it's in the combustion chamber. If you have a wood pile, it'll burn very hot and fast if you're not careful. Always consider fuel storage while you're designing your stove system.

✔ Always make sure to have adequate smoke alarm coverage in your home, and make sure to keep the batteries fresh in every smoke alarm. You should also consider installing a carbon monoxide alarm. It's always better to be safe than sorry.

Chapter 15

Hydrogen Fuel Cell Technologies

*H*ydrogen is the simplest and most abundant element in the universe. It is comprised of a single proton, and hydrogen gas is odorless, colorless, and imperceptible to the human senses. It's the main component of water (H_2O), and it's also one of the basic atoms in both carbohydrates and hydrocarbons, the energy sources that all life on the planet earth require to survive.

It is no stretch to say that hydrogen is the most important element in the universe because the energy from stars derives mostly from it. It's also the most important element in terms of the world's energy future, for it's also no stretch to say that sometime in the future humankind's main energy source will be hydrogen. This is going to take some time, but it's a fact simply because it has to be. We must develop a hydrogen economy or perish. Fossil fuels are going to run out, and greenhouse gases from fossil fuel consumption are causing too much environmental harm. Hydrogen provides the answer to both dilemmas.

In this chapter I describe hydrogen fuel cells, the basic technology at the heart of the hydrogen promise, and explain why hydrogen is so important in the grand energy scheme.

The Basics of Hydrogen Technology

In the last few decades, fuel cell technology has developed and the most widely used type of cell is the hydrogen fuel cell. Basically, a hydrogen fuel cell produces electrical power from hydrogen, which combines with oxygen in a specially manufactured device. Hydrogen and oxygen (the raw fuels) are fed into the device, and electrical power is output (along with heat and water and some other minor byproducts).

Instead of combusting as hydrogen does in hydrocarbons (fossil fuels), the hydrogen oxidizes at a much lower temperature. In the most common scheme, a *proton exchange membrane* (PEM) is employed. A PEM fuel cell generates between 0.4 and 0.8 volts of electricity. In the same way as batteries, individual cells are combined in order to create higher voltages, or higher power levels. *Note:* Don't worry about how a PEM works: it's very complicated, would take an entire book to explain, and isn't necessary for this discussion.

Hydrogen is not really an energy source, per se; pure hydrogen offers no potential to extract useable energy. But it's perfect as an energy carrier (storage mechanism) because it's ideal for converting energy from other sources, and converting that energy directly into electricity. Herein lies its magic. The combustion process associated with most other forms of energy extraction is bypassed. With hydrogen, pure electricity can be produced without need for boilers, turbines, internal combustion, and generators. The process is very simple, and simplicity implies efficiency.

Sources of hydrogen

Hydrogen is very reactive with a lot of other atoms, so it's rare to find pure hydrogen in a natural state. In order to use it as a fuel, therefore, it must be separated from other chemical compounds, of which there are many widely available candidates, the main ones being

- **Water:** Water is H_2O, as everyone knows. When water is broken down to produce hydrogen, the only other product is oxygen, which is completely beneficial. Contrast this with the production of hydrogen from hydrocarbons, which releases a great deal of carbon in the process, and it's easy to see why water is the best source of hydrogen.

- **Hydrocarbons (fossil fuels):** While it may seem contradictory to extract hydrogen from fossil fuels for use in producing electricity via fuel cells, the process is far more efficient than the current combustion processes of fossil fuels. So the net gain in terms of both pollution mitigation and energy efficiency is very promising. Still, environmentalists object to fuel cell development that uses hydrocarbon sources (such as propane) because it does not solve our energy problems nearly as effectively as renewable hydrogen sources such as biomass.

- **Carbohydrates (food and biomass products):** Biomass methane is an environmentally sound source for hydrogen, and the biomass industry will likely grow in conjunction with fuel cell proliferation. (Refer to Chapter 13 to find out more about biomass.)

Nearly any material that reacts with oxygen may be used, but efficiencies and economics vary widely, and most substances that can produce hydrogen are impractical due to cost and complexity.

Globally, a well established hydrogen industry currently extracts 50 million tons of hydrogen per year. In the U.S., the oil refinement industry uses around 50 percent of this hydrogen in their processes, and fossil fuels are the main source of the hydrogen. Natural gas provides 48 percent of the hydrogen, oil provides around 30 percent, and coal another 18 percent. This is not an environmentally friendly mix of sources.

Hydrogen is so prevalent that there are a wide range of sources from which to derive it. Yet pure hydrogen is a difficult fuel to obtain. While many scientists and engineers believe that hydrogen fuel cells are the key to the world's energy future, there is still considerable debate about the nature of the sources from which the hydrogen is derived. Hydrogen from non-renewable sources carries the same burdens as our current use of these resources. When acquired from clean, renewable sources, however, hydrogen can offer a virtually limitless abundance of electrical energy with nearly negligible impact on the environment. The key to a hydrogen future will be to develop the right sources, and the means to separate the hydrogen from those sources.

Extracting hydrogen

As mentioned previously, hydrogen can be extracted from nearly any hydrocarbon as well as carbohydrate. The following sections outline some of the most common methods of hydrogen extraction. Research and development is being funded for all of these, and at some point there will be technical breakthroughs that favor one of the methods over the others.

Steam reforming

The most widely used process, called steam reforming, converts natural gas into hydrogen. Steam reforming combines methane and other lightweight hydrocarbons with steam (water vapor at a specific temperature) causing the constituent chemicals to react. The products are hydrogen, carbon dioxide, carbon monoxide, water, and a number of other chemicals that cause pollution hazards. The hydrogen is separated and stored.

Coal based steam reforming

Coal may be converted into a gaseous form that combines coal, oxygen, and steam under tremendous pressures and temperatures. A gas of carbon monoxide and hydrogen forms, and the hydrogen is separated out into pure hydrogen raw fuel.

Plasma waste

A company called Startech has demonstrated a new technology that can produce hydrogen from trash. The process is called plasma waste remediation and recycling technology. Waste materials are fed into the device and reformed and recovered by molecular dissociation and a closed-loop elemental recycling process. Economical it's not, but the technology is sound. With improvements, this may be the wave of the future.

Here's how the process works. In an ambient of ionized air, waste materials are jolted hard with energy to the point where it causes molecules to break apart into actual atoms. The byproduct is *plasma converted gas* (PCG) from which hydrogen can be easily extracted. Startech has converted a conventional pickup truck that runs off the technology.

Electrolysis

Hydrogen can also be produced from water, either salt or fresh, through electrolysis, and this is highly favored because there are no greenhouse gases as a consequence. Electrolysis divides water molecules into basic elements of hydrogen and oxygen (somewhat the opposite of how a fuel cell works). A virtually limitless supply of hydrogen is produced if renewable sources such as water are used as the basis. But electrolysis itself requires electricity, so this process is pricey compared to steam reforming. It is estimated that electricity costs need to be less than 3 cents per kWh for this process to be economically competitive.

Biomass gasification

Biomass gasification uses hot steam and oxygen to convert biomass into a gas, which is a mixture of carbon monoxide, hydrogen, carbon dioxide, water, and other byproducts. The hydrogen is separated and used as a fuel source.

Thermal dissociation

Thermal dissociation is still on the drawing boards, but it shows promise. Thermal dissociation requires extreme temperatures (3,600 degrees Fahrenheit or more) which are obtained by concentrating solar systems. The process separates hydrogen and oxygen in water thermally, and is very effective. Research is being done to find a way to lower the temperatures required, and if successful this may be the major hydrogen producing method in the future.

How a fuel cell works

There are a wide variety of fuel cells, each using a similar process to produce electricity. A fuel cell uses an electrochemical reaction to convert hydrogen and oxygen into electricity, water, and heat. The process transforms fuel directly into electricity in a single step. This single step makes fuel cells very efficient in terms of converting potential energy into useable electrical energy.

A fuel cell has two electrodes, or electrical connections: a cathode and an anode, with polarities. The two surfaces are separated by a membrane electrolyte, a liquid solution with carefully controlled electrical conduction properties. Hydrogen gas is channeled into the anode side of the cell. A catalyst, typically platinum metal, oxidizes the hydrogen and this results in the separation of the hydrogen proton and its single electron. The electron flows out of the cell via the electrodes and an electrical current is generated.

At the same time, the now positively charged hydrogen ion flows through the membrane to the cathode (due to electrical attraction). At the cathode, air is electrocatalytically transformed into negatively charged oxygen atoms. The force of electrical attraction now bonds the positively charged hydrogen ion with the negatively charged oxygen atom, creating water in the process. The final product is water, heat, electricity, and some nitrogen oxide in very small amounts. The water may be simply discarded, the heat channeled away and used to warm external devices, if desired. It's the electricity generation that's of prime interest.

Whew. That was hypertechnical, but don't worry, you don't need to understand that a fuel cell operates just like a battery. It outputs DC (direct current) power to a load. The load may convert this into AC electrical power suitable for residential and commercial use.

The low temperature operation of the fuel cell is a function of the technical precision and complexity of the catalysis reaction on the fuel cell catalyst (membrane). While traditional thermal engines rely on temperature differentials to achieve the efficient production of useable energy, fuel cells operate differently. With the proper catalysts, the hydrogen and oxygen is "burned" atom by atom right on the surface of the catalyst itself to produce not heat but a physical entity even "hotter" — a flow of electrons. Fuel cells do not violate the thermodynamic laws (refer to Chapter 2), they just achieve them in subtly different ways than thermal combustion processes, which are far more intuitive to the layman.

Back story in brief

The Welsh physicist Sir William Robert Grove built the first prototype fuel cell in 1839, and demonstrated that hydrogen and oxygen, under the proper conditions, can be made to react to create an electrical flow (with only water and heat as a byproduct). This first practical application of fuel cell technology was very costly compared to other forms of energy production, but it proved the viability of the fundamental concept.

In the late 19th century William Jaques built a fuel cell that used phosphoric acid as the primary fuel. This embodiment offered even better performance than Grove's invention. Solid oxide fuel cells (SOFC) were developed by German researchers over the course of the 1920s. In 1932, Francis Bacon, an English scientist, used alkaline and nickel to significantly improve fuel cell performance. This technology took hold, and was improved by subsequent researchers and a 5 kilowatt system was built in 1959 that demonstrated the practicality and economical potential of fuel cells.

Not until the American space program in the 1960s were fuel cells used in working systems. Fuel cells are lighter weight than batteries, very compact in terms of output energy per size, and much safer than nuclear power reactors, so many spacecraft were outfitted with them. In 1965, Project Gemini utilized fuel cells to provide heat, electricity, and potable water supplies on board a working spacecraft. All space shuttles are now outfitted with fuel cells. The economics are still prohibitive, but the advantages in space applications merit the high cost. But by developing fuel cell technologies, insights and new innovations have been made and the costs have been coming down.

Practical and Pie-in-the-Sky Applications

Given the fact that fuel cells create electricity directly, there are very few energy applications where fuel cells are not great candidates to displace current technologies. Hydrogen, unlike electricity, is simple to store. And oxygen can be obtained from the air, so there is no problem with that supply. Hydrogen can be compressed down into gas tanks, then combined with oxygen from the air and electricity may be generated at will. Wherever batteries are used, hydrogen fuel cells may be used.

✔ Rural homes that are off-grid will likely be the first big market for fuel cells.

✔ Portable devices that use batteries are also prime candidates for fuel cells. Devices such as smoke alarms, video cameras, and cell phones will likely be available with fuel cells by 2012.

✔ Electric cars will someday use fuel cells, and the only exhaust will be water and some minor gases (non-greenhouse). At this point, there are over 3,000 fuel cell units being used in institutions and businesses worldwide.

✔ All the big automakers are exploring fuel cells. Projections are for the first fuel cell passenger cars to appear on the road in 2010, at the earliest.

The following sections discuss two different and simple applications for fuel cells, to give you an idea of how useful they are, and one use that gives you a glimpse of a hydrogen-powered future.

Fuel cell-powered vehicles

A fuel cell-powered vehicle (FCV) is basically an electric car that uses fuel cells as the primary power source. In such a car, hydrogen and oxygen are fed into the fuel cell, which produces electricity which is directed into the battery bank (conventional batteries are used) and the power controller. The battery bank is used to allow for varying amounts of power to be directed to the electric motor (fuel cells are poor at decreasing and increasing their power outputs at a fast enough rate to directly power the wheels, although this is changing with new technological developments).

The electric motor is connected to the drive wheels via a transmission (as it is in conventional vehicles). The power controller is connected to the accelerator pedal that the driver manipulates to control the vehicle speed. The battery also provides another benefit: If the fuel tank (hydrogen) goes dry, there will still be available power to get to the nearest refilling station.

With current technologies, fuel cells can convert around 50 percent of the potential energy in the hydrogen into useable electrical energy to power the vehicle. The rest of the energy transforms into heat, which is dissipated by an onboard cooling system (just like conventional autos that use radiators to cool the engine).

Following are the advantages, disadvantages, and challenges regarding broader use of FCVs:

✔ **Efficiencies:** FCVs help reduce dependence on foreign supplies of fossil fuels. Even if the hydrogen source is a fossil fuel, fuel cells offer far better efficiencies than conventional engines so demand for fossil fuels will be less when FCVs become commonly available. In addition, electric vehicles of all kinds are lighter weight than conventional vehicles because there is no need for heavy, mechanical drivetrains. This further enhances the efficiencies.

- ✔ **Infrastructure:** Before FCVs become common, there will have to be a lot of investment in infrastructure like special fueling stations and the transport of raw fuel sources to and fro. Fortunately, existing natural gas and methane pipelines can be used to channel hydrogen gas, so some infrastructure is already in existence, even though it will have to be modified.

- ✔ **Environmental impact:** The pollution generated by an FCV is a fraction of that exhausted by a conventional vehicle. When biomass is used as the source of hydrogen, the emissions are closed loop (I explain this phenomenon in Chapter 13).

- ✔ **Cost:** The current cost of hydrogen is currently very high, and availability is virtually nil. When hydrogen production facilities come on line for transportation, economies of scale will result in lower prices and hydrogen fuel cells will become more widely used as a result. The growth process will feed on itself. Government incentives are very encouraging for this technology, so the costs are being kept low in order to spur demand. Government incentives are growing every year.

The transport sector is going to drive the demand for fuel cells, and open up new residential and commercial building markets as prices decline.

- ✔ **Transportation of fuel:** Hydrogen can be produced locally and doesn't need to be manufactured in large industrial plants far removed from filling stations. Hydrogen can be produced at a residential level which means that it will be possible to completely forego filling stations.

- ✔ **Operating range:** The effective operating range of an FCV with a full tank of fuel is less than that for a fossil fuel-powered vehicle.

- ✔ **Fuel tanks and car design:** Onboard storage of fuel is problematic. Hydrogen tanks are difficult to build because the atoms are so small they permeate most metals. Fuel tanks will always be expensive and difficult to maintain. In addition, these tanks are larger and more complex than fossil fuel tanks and take up more space; since electrical vehicles are smaller and lighter weight, the interiors of FCVs are more cramped.

- ✔ **Maintenance challenges:** There are very few FCVs on the road now, and maintenance will be difficult because most mechanics have no idea how an FCV power source works. It is going to take not just a considerable investment in infrastructure, but also a considerable investment in training and licensing of qualified service personnel.

Small-scale fuel cell electric power

Electrical power may be produced on site, using fuel cells. Residences and businesses can produce their own electrical power. *Small scale* basically means that the power output is low and is only capable of driving a limited amount of load. Typical small-scale power plants can produce anywhere between 5 to 50 kilowatts of power. The advantage is that the power is produced independent of the grid, which makes for a more distributed form of energy production.

In a small-scale fixed location system, hydrogen can either be piped in via the existing natural gas pipeline system or stored on site in a tank. (In fact, if methane — natural gas — is used to provide the hydrogen, the fuel supply is already in existence in most urban areas.) Another alternative is to produce the hydrogen on site with a reformer or other type of hydrogen recovery system, explained in the earlier section "Extracting hydrogen."

The hydrogen and oxygen are then combined to create electricity, which feeds into the power control circuit. When load conditions in the home are minor, most of this power is fed into the batteries where energy is stored for later use. When the load conditions in the home are heavy, most or all of the power being generated by the fuel cell goes directly into the home's appliances. If the hydrogen fuel tank runs out, there will be some energy capacity stored in the batteries.

Fuel cells are very simple, in comparison to other energy generating schemes used in residences, and using them to generate electricity on a small scale offers several advantages:

✔ The cells last a long time and are very reliable. For remote locations, they're more efficient than conventional generators used as backups and primary sources of power.

✔ Hydrogen is non-toxic, unlike gaseous fossil fuel energy sources, and there is negligible pollution produced in the process. Even if conventional fossil fuel sources such as methane and propane are used as fuels to the fuel cell, less pollution results than if those conventional sources were combusted on site to create electricity. In fuel cells, there is no combustion at all, and that's the difference.

✔ Tax rebates and credits and subsidies are available for these systems, which lowers the net cost.

Despite the advantages, there's one key disadvantage currently (beyond the fact that the technology is very rare and getting units serviced may be problematic, to put it mildly): It's very hard to deliver and store hydrogen. The energy density of hydrogen is low compared to other fuels, so storage tanks need to be larger to contain the same amount of potential energy, and Hydrogen is extremely flammable and potentially explosive. On-site production will be the key to the future proliferation of these technologies, but it still has a long way to go.

Hydrogen-powered home of the future

At some point in the future, hydrogen fuel cell problems will be solved, and the resulting growth of technologies will be tremendous in both the residential and transport sector. Figure 15-1 shows an artist's conception (mine) of a hydrogen-powered home of the future.

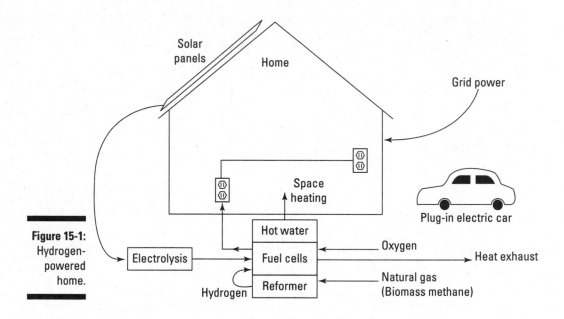

Figure 15-1:
Hydrogen-powered home.

At the core of the energy process will be hydrogen-powered fuel cells. Hydrogen will be obtained through a range of sources, like biomass and natural gas, for example, with natural gas the most likely. Hydrogen can also be produced via electrolysis which is driven by solar panels, which will completely eliminate greenhouse gas emissions. As solar power comes down in cost over the next decade, it will become a more economical alternative to natural gas hydrogen production.

The heat from the fuel cell electrical generation will be used to heat the domestic water supply and the internal spaces of the home. This makes the process extremely efficient.

The home will still be connected to the grid, but with an adequate size of fuel cells and solar panels, there will be a net zero transfer of electrical power. The grid will ensure that if the fuel cell system goes down there will still be electrical power available. Another attractive feature of being connected to the grid is that fuel cells and solar panels can output lots of power during peak times. This will increase the efficiencies of the entire grid system (I explain this in more detail in Chapter 3).

Finally, plug-in electric cars can be recharged via the electricity produced by the fuel cells. Recharging will occur mostly at night, but at this time the home is using minimal power so the load is spread out.

The Future of Fuel Cells

Since first being used to successfully drive a transport vehicle (a bus) in 1997, a whole host of new applications have been achieved for fuel cells, as the earlier section mentioned. At this point, fuel cell viability has been amply demonstrated. The problem is that hydrogen fuel cells still cost too much, but it's only a matter of time before fuel cells become competitive with common energy sources.

Currently, hydrogen fuel cells as an energy source are rare because of their high cost, only being found in research institutions and university engineering labs. But much progress is being made, and the promise is so outstandingly bright that a lot of new research money is being directed toward developing hydrogen energy schemes. The growth in viable hydrogen technology is increasing geometrically.

Despite the fact that fuel cells in general are pie-in-the-sky at this point in time, their importance is critical to understanding alternative energy sources. Even if the proponents' wildly optimistic projections fall short, fuel cells are going to be ubiquitous within the next decade. Most likely, hydrogen will play a major role, but will never completely displace other forms of energy production. But even if that's true, the world will be much better off.

Still, a number of hurdles must be crossed for hydrogen fuel cells to become commercially viable:

✔ The lifetimes and reliability of fuel cells must be improved. Despite their relative simplicity, fuel cells wear out just like batteries, and in light of their high price, are very costly to replace.

✔ Initially, fuel cell vehicles are going to cost more than $100,000, which is around four times the typical price of a new car with similar performance characteristics. While people are willing to pay more for green technologies, this 4 to 1 ratio must come down to something on the order of 1.5 to 1 before widespread market acceptance occurs.

✔ Hydrogen fuel prices need to come down. Extraction methods need to be developed which put the cost of raw hydrogen fuel on a par with fossil fuel competitors. However, if cap and trade policies and carbon taxes are implemented, the gap in raw fuel costs will be decreased by a big margin since hydrogen is such an environmentally benign source of energy.

✔ Hydrogen fuels obtained from fossil fuels are by far the cheapest option, at present. This does not solve our energy problems, and in light of the inefficiencies inherent in extracting hydrogen may even make matters worse. New extraction techniques that work with biomass need to be developed and perfected, and even then, there are not enough biofuels to satisfy the demand for hydrogen, if it's anywhere near what projections indicate it may be.

✔ Hydrogen gas is easily ignitable, and poses inherent dangers. Storage means must be developed which are reliable and safe. But the fact is, when hydrogen burns it burns very quickly, and the only product is water. Combusting hydrogen emits 90 percent less heat than combusting fossil fuels, so the danger of being burned is much smaller. But this is a tough sell to consumers who only hear the word "danger."

✔ Transport of hydrogen gas can be problematic as hydrogen is a very small molecule that tends to leak through containers. And since humans have no sensory perception of hydrogen gas, there is no detecting leaks. Keep in mind that the Hindenburg airship caught on fire in 1937 and 35 of the passengers died. There were movies of this accident shown worldwide, and many people still associate hydrogen with the Hindenburg. Yes, it's the same stuff.

✔ Finding an effective way to deliver hydrogen to the fuel cells used in the transport sector is also becoming a show stopper. It will take considerable investment in expensive infrastructure to provide enough hydrogen filling stations to interest people in buying very expensive new cars that run on hydrogen.

✔ Compressed hydrogen gas has a low density and so it has a low energy-to-volume ratio as well. Vehicles that use hydrogen have larger tanks, and travel less distance between fill ups. Advances in storage will need to be made in order to solve this problem.

✔ An alternative to carrying onboard hydrogen gas is to incorporate onboard partial-oxidation reformers into the vehicle which convert fossil fuels into hydrogen fuel. The technologies are immature at this point, but this may be the most promising solution.

Part IV
Alternatives — Transportation

The 5th Wave   By Rich Tennant

My car runs part of the time on an alternative energy source.

In fact, here it is now.

KERM'S TOWING

In this part . . .

*V*arious alternative energy schemes are being developed for the transport sector of the economy. A number of alternative fuels are already available, including biomass-produced methane, propane, and biofuels. The first cars were all-electric, and in Chapter 18, I describe the advantages (and disadvantages) of electric cars, which are seeing a major resurgence. In Chapter 19, I describe how hybrid vehicles work and why they make so much sense. Chapter 20 addresses fuel cell–powered transport, which many believe (perhaps prematurely) is going to be key to a fossil fuel–independent future. In Chapter 21, I present some of the more promising exotic transportation options, such as magnetic levitation trains and nuclear-powered cargo and passenger ships.

Chapter 16

Alternative Transport Technologies

- -

In This Chapter

▶ Exploring a brief history of transport technologies

▶ Following the relatively recent evolution of the internal-combustion engine

▶ Checking out what the future holds

- -

Currently, the internal-combustion engine (ICE) powers the vast majority of vehicles, but this wasn't always the case. Some of the earliest automobiles were powered by electricity. The future of transport harkens back to this pre-ICE era. Getting there will take considerable investment in both infrastructure and ingenuity, but the baseline has already been established, by the earliest car designers, no less. How's that for a little technological irony?

In this chapter I describe how transport power has evolved, how current advancements are driving new technologies, and why all-electric transport is a good thing.

Turning the Pages of Alternative Vehicle History

At the turn of the last century, steam power, fueled by coal or wood, was all the rage. It powered factories and locomotives, all the while belching thick, black smoke clouds (borne of a basically inefficient technology). Because steam engines were a mature technology at the time, it only made sense that they be used for cars. Internal-combustion engines, still in their infancy, required a lot more complexity and support equipment than a simple steam engine. But steam had its own disadvantages, which fueled the search for alternatives. The following sections explore the scope of alternative vehicles throughout history from steam-powered to electric to hybrids.

The glory days of steam

In 1825, the Goldsworthy Gurney company in England built a practical, steam-powered car that could travel over 80 miles in ten hours. That's 8 miles per hour, which is not much faster than a horse. Fuel was prevalent however. If you ran out, all you had to do was take a hatchet into the woods and chop away for a little while.

In 1897, steam-powered transport was more prevalent than gasoline-powered transport simply because internal-combustion engines were so much more complicated. And although electricity was the most favored form of power (because you could simply hop into an electric car, press the accelerator, and be on your way), steam power was the most common form of auto propulsion in the United States. Manufacturers sold 1,681 steam cars that year; 1,571 electric cars; and only 930 gasoline-powered cars.

But steam was not without its problems. It took a considerable amount of time for the steam boiler to get up to temperature. You could not simply jump into a steam-powered auto and take off. You had to start a fire and wait until the raw fuels were burning hot enough to boil water. Then you had to continuously feed the fire with new fuel and remove the cinders and ash products that were left behind in the combustion process. Plus, if you were anywhere near the engine, it was a rather hot venture.

Electric-powered vehicles

The first electric vehicle was built in 1839 by a Robert Anderson. Electric motors are very compact and dense compared to gasoline engines, and they run silently — all good things. The main problem with electric cars is the batteries.

Back in the day, batteries weren't rechargeable. When the batteries ran out, new ones had to be installed. Plus, batteries weigh a lot and are made of noxious chemicals that can cause dangers if used improperly.

Battery technology is still the main limitation of electric-powered vehicles. Despite new high-tech battery technologies, they still weigh a lot and are still made of expensive, potentially dangerous chemicals.

Despite these problems, electric vehicles of all kinds proliferated during the first couple of decades of the 20th century, beating out gasoline-powered autos. The reason for the popularity of early electric cars was that they were much simpler, lighter weight, easier to fix, and safer — attributes that all sub-sequent versions of the electric car can also claim. Because of these advan-tages electric cars have always been just around the corner. They probably would've rounded the bend already if not for the darn batteries. Battery tech-nology has been steadily creeping forward, and with the burgeoning concern for the environment, there is even more impetus to develop electric cars. In Chapter 18, I describe the newest electric car technologies. Battery issues aside, while there is still a ways to go, electric cars are now competitive eco-nomically and technologically with fuel-driven cars, and they're a lot cleaner to boot.

Hybrids enter the scene

In an attempt to solve the battery problems associated with all-electric cars, hybrid vehicles were developed. The first hybrids used gasoline engines only to generate electricity, which was then fed into an electrical motor which directly powered the wheels. A number of versions of this basic design were developed, none of them gaining widespread acceptance.

But with advancements in battery technologies, and with increased con-cern about environmental harm from fossil fuel combustion, hybrids have become one of the best alternatives to the internal-combustion engine. Hybrids feature gas mileage figures over 50 percent better than equivalent gasoline-driven autos, and the lifetimes of the battery banks have increased from a couple years to over ten years. Some auto manufacturers are war-ranting hybrid battery banks for the lifetime of the car. For more on the design, technology, and characteristics of hybrid cars, head to Chapter 19.

Leavin' on a jet . . . car?

Gas turbines (See Chapter 5 for more details on turbines) were also used in early auto technolo-gies. A turbine is closely related to a jet engine, although with a jet engine it's the high-velocity exhaust gases that move the plane. When a tur-bine is used in an auto, the power is channeled directly to the transmission, which drives the wheels. Turbines can run on a variety of fuels with equal effectiveness, and this was seen as a major benefit.

Due to some initial successes it was believed that all cars would eventually be powered by turbines. But turbines were notorious gas pigs. Efficiency was terrible and the sound was enough to break eardrums. Contrast this with the silence of electric cars and it's easy to see why turbines died out for use in vehicle transport.

Evolving from Mechanical to Electrical

Each of the mechanical parts of a conventionally-powered internal-combustion engine (ICE) weighs a lot, and takes up a lot of room as well. (For a complete discussion of how internal-combustion engines work, both diesel versions and gasoline versions, head to Chapter 5.) Mechanical parts wear from friction, and there is a lot of leakage of lubricants (called evaporative emissions, a source of air pollution). There are literally hundreds of parts that need to be changed out, like belts, pumps, and fans. Collectively, the drivetrain parts of a typical conventional auto comprise around half of the total vehicle weight.

To address inefficiencies associated with internal-combustion engines, some significant changes have occurred in the last couple of decades, all made possible by the advancements of silicon-based microprocessors. The following sections take you along for the ride from a more mechanically based ICE to an electrical one.

Watching the rise of silicon

Silicon-based microprocessors have revolutionized the world over the last few decades. From year to year, computer technology improves by leaps and bounds. With these microprocessors, computers process huge amounts of information efficiently and quickly.

But silicon control of high power, meaning hundreds of horsepower, has lagged behind. This is now changing, and it's possible to control automobile-sized scales of power via silicon transistors. Due to these advances in silicon power control technologies, the internal-combustion engine has been changing radically in the last two decades.

Every year, more and more of a typical car's powertrain will be converted from mechanical to electrical, controlled principally by digital computers. This is made possible by the improved technology of silicon power control chips. Silicon power chips, like their micro-sized brethren used to power a computer, have been improving drastically over the last few decades. They're getting better and cheaper all the time, and this provides an excellent opportunity to make cars cheaper and more efficient as well.

Decreasing demand for mechanical controls

Internal-combustion logic is still the same as it was back in the days of the steam engine: Valves click and snap against each other. Friction consumes a large portion of an engine's overhead (that is, it takes a lot of power simply to run an engine, which is inefficient), and the peripheral control mechanisms surrounding an engine consume around the same weight as the engine itself. Radiators full of coolant, for example, are required to maintain an engine at very specific operating temperatures. Generators provide the car's electrical systems with power and these take even more weight. Elaborate timing mechanisms operate belts and pulleys and valves, all of which rub against each other and wear out.

Compared to the elegance and simplicity of modern microprocessors, an internal-combustion engine resembles a Rube Goldberg contraption, and the entire process is extremely inefficient. In fact, a coal-fired electrical power plant is twice as efficient at converting raw fossil fuels into useable power than an internal-combustion engine is.

The trend toward more electrical control is very strong because with better valve controls and better fuel sensors, new engines are far more efficient than older versions.

While the first cars were electrical and quite simple in both construction and operation, today's autos feature over 5,000 feet of copper wire with hundreds of connectors, fuses, and relays to transmit the electrical power where it is needed. Electrical demand in a modern auto is increasing by 5 percent per year, whereas the mechanical demands are decreasing due to better designs and improved efficiencies.

The cost of a new car is disproportionately heading toward electrical computers over gears and mechanical devices. In fact, Detroit now spends more on etched silicon control devices than on steel.

Getting efficient: Same power, less weight

Since humans first began riding horses for transportation there has been approximately one ton of vehicle weight to move one human passenger. This ratio has been consistent through the development of automobiles and passenger jets. A ten-to-one vehicle-to-passenger ratio means that the energy efficiency ratio is also ten-to-one.

Most of an automobile's weight is dedicated to the engine, transmission, and drivetrain. Masses of steel are required to ensure the integrity necessary to support the mini-explosions caused by compressing the fuel-air mixture within engine cylinders. Heavy gear boxes are required to enable an ICE to operate at its most efficient levels. All this weight increases inefficiency because most of the energy expended on transportation is basically wasted on transporting the transportation, as opposed to the passengers and cargo. Therefore, making engines lightweight and using very energy-dense fuels can reduce these inefficiencies.

As electrical power controls become more refined and intelligent, the mechanical powertrains will give way to electric motors, which weigh far less and take up much less space. Electric motors can be connected to the wheels, without need for complex transmissions and gear boxes.

Recognizing the car of the future

For an auto of the future, the starter motor of a conventional auto is now replaced by a dual-purpose starter/alternator. Battery power is used to crank the starter, and the ICE comes on, at which point the starter/alternator switches into alternator mode and hundreds of kilowatts of electrical power are produced. This electrical power is then used to power all the onboard functional devices.

Each wheel on a car can have its own electric drive motor, and the net weight of four electrical motors will be much less than the net weight of all the gears and transmissions parts that they displace in a conventional car. The ride will be better as well. Cars will shed literally hundreds of pounds of weight without sacrificing an ounce of performance.

This car design is far more efficient than a conventionally-powered ICE because the car will weigh so much less, and electrical power can be precisely controlled by an onboard microprocessor to channel power only when and where it is needed.

If this seems futuristic, the General Electric AC6000CW locomotive is now powered by huge diesel-fueled (efficient and clean burning; see Chapter 17 for more on diesel fuels) electrical generators. The entire machine is electric, aside from the diesel engine which does nothing but turn a generator that produces electrical power. Large trucks are also on the road that run the same way: The diesel engine simply turns a big generator and all the motive force is provided by electric motors coupled to the wheels.

Electrical load in modern cars

The total electrical load in a car is now about 1 kilowatt, which is a lot of power. Peak loads exceed 2 kilowatts (air conditioners and heaters take a lot of power, as do big stereos). By contrast, the mechanical power plant in an ICE vehicle outputs around 100 kilowatts, with an average output of 20 kilowatts.

Military platforms like jet planes and submarines are all electric, as are aircraft carriers. The Segway, an 80-pound, dual-wheel scooter is entirely electric. The performance of each of these examples exceeds that of their conventional counterparts.

One problem with this scheme is the unreliability of digital controls. You have surely experienced your computer crashing. You need to turn it off and then reboot it to get it working again. With too much computerized intelligence onboard a vehicle, there is a possibility that the control system will crash. Much development is being directed at this problem, and the most promising solution is to use redundant computer controllers so that when one goes down, others are still on line. With the low cost of computers, this does not add a lot to the total cost of the vehicle.

Coming Full Circle — and Back to Electric

Over the last century, internal-combustion engines are supported by a massive infrastructure and intellectual property in the technologies. Companies have spent untold millions developing new ideas and machines, and when an industry changes, all those investment dollars are wasted and new investment dollars need to be spent on the new technologies. Companies don't like to waste their money, so they tend to resist new technologies, not because they're inferior but because it's uneconomical. In addition every corner gas station has a mechanic that can recite the details of internal-combustion operation. It will take a considerable investment in new infrastructure before all-electric vehicles displace internal-combustion vehicles.

Whereas the gasoline-powered vehicle was the alternative of yesteryear, supplanting steam- and electricity-powered vehicles, today the tables have turned, and electric-powered vehicles are making a comeback. After several years being transported by gasoline-powered engines, cars are once again returning to their electrical roots and, in the process, becoming little more than large electrical appliances — a trend that will continue to grow over the next few decades. The goal? To completely displace the internal-combustion engine, and with it all the fossil fuel problems inherent to internal combustion.

Barring a complete turn away from internal-combustion engines, the other option is to create alternative fuels that make internal-combustion engines more efficient. See most of the remaining chapters in this part — Chapters 17 through 20 — to examine all these other alternatives.

Chapter 17

Alternative-Fuel Vehicles

A number of alternative fuels for vehicles have been developed and tested to replace the conventional gasoline and diesel energy sources. The goal is to use renewable, clean sources of energy, and to effect better efficiencies.

This chapter explains three of the main alternative fuels: natural gas, flex fuels, and biodiesel.

Natural Gas in the Tank

Although natural gas (in gaseous form) is commonly associated with home and commercial heating, it can also be used to power a vehicle. Nowadays, natural gas is used to power a number of buses and garbage trucks, but the expenses are higher than for conventional gasoline-powered vehicles because the technology is still rare. But this is changing. The following sections explore the advantages as well as the practical applications of natural gas as an alternative for vehicle fuel.

Around the turn of the last century, natural gas was used before gasoline in early vehicles. Because gasoline is easier to transport and pump into a tank, when the first gasoline-powered vehicles hit the road, the use of natural gas quickly diminished.

Advantages to natural gas

With the new interest in renewable and domestic energy supplies, natural gas is once again being considered as a fuel source for vehicles. Natural gas can be made of many different biomass sources, as explained in Chapter 13. The key environmental advantage is that the carbon emitted by biomass-produced natural gas is a closed loop. When fossil fuels are drilled from the ground, the carbon released by combustions has been sequestered deep within the earth for millions of years. By releasing this carbon into the atmosphere, the net amount of carbon in the atmosphere increases. When biomass sources are used to produce natural gas, the carbon merely changes forms, so there is no net increase of carbon in the atmosphere.

Perhaps the biggest advantage of natural gas propulsion is the fact that it relieves our dependence on foreign oil supplies. Eighty-five percent of the natural gas used in this country is domestically produced. Contrast that with less than 45 percent of the crude oil that is consumed here. Furthermore, of all the natural gas that America does import, most of it comes from Canada, which has a friendly, stable political system. Even then, imports are rising, and most of the current natural gas reserves are in Russia, which may be an even more problematic source of raw fuels than the Mideast.

Producing natural gas

The most common variety of natural gas is methane, or CH_4 (four hydrogen atoms, one carbon atom). Waste materials like compost from farms is converted into methane gas, which produces both fertilizers and fuel suitable for use in a natural gas vehicle. The gas is fed into a pipeline distribution system or into large holding tanks. The gas may be used for both home heating and vehicle propulsion.

Natural gas production is local. It can be produced on farms, or in small quantities at special plants dedicated to the purpose. The plants are not particularly expensive, so natural gas is price competitive to regular gasoline.

One quick note, that you may find interesting, alarming, or just a tad of both: The production of natural gas from biomass may lead to the spread of nasty little microorganisms and pathogens. This is the stuff of science-fiction movies, but it's not out of the realm of possibility.

Natural gas vehicles

Natural gas vehicles incorporate internal-combustion engines (ICEs) that have been modified to work with the natural gas. Converting conventional engines to natural gas is relatively easy; a skilled mechanic can do it in a weekend. The United States currently has over 150,000 natural gas vehicles. Before you start converting all of the vehicles in your garage to natural gas vehicles, check out the considerations in the following sections to be sure you know all the facts.

Exhaust emissions

The exhaust emissions are much less than with conventional engines. The Department of Energy (DOE) estimates that natural gas vehicles emit 90 percent less carbon monoxide and 60 percent less nitrous oxides. In addition, natural gas engines emit around 35 percent less carbon dioxide. Natural gas-powered engines are more efficient at converting the potential energy of the raw fuel into motive force for the vehicle. In addition to less emission per unit of raw fuel, therefore, natural gas vehicles get better gas mileage, too.

Performance advantages

First the good news: A well-maintained natural gas vehicle will last longer than a conventional vehicle. Because natural gas burns cleaner than conventional gasoline, there are fewer combustion byproducts that can blow the piston rings and contaminate the lubricants used by the engine. No gasoline filters are required and the fuel injectors don't jam the same way that conventionally-powered vehicles do. In addition, the cost of natural gas fuel is around half that of gasoline, which translates into a much smaller per mile cost to drive.

Performance limitations

Now the not-so-good news: Natural gas-powered vehicles, while more efficient than conventional fossil fuel vehicles, are limited in terms of the power they can produce. Natural gas-powered cars are slow off the mark, and can have problems accelerating up hills.

Natural gas is also more difficult to deal with than liquids: It must be compressed to a pressure of around 3,000 pounds per square inch, so tanks and containers must have a lot of integrity and meet special safety requirements and undergo periodic inspections by licensed professionals. These tanks are also heavier and larger than conventional gasoline tanks, which makes for a more crowded vehicle. In addition, you have to constantly monitor the natural gas tank to make sure there are no leaks and that the tank is maintaining its pressure.

The one real practical limitation is that you cannot take a natural gas vehicle cross country because you simply can't get it filled in most places (see the next section "Availability of natural gas for vehicles" for more information). Therefore, natural gas vehicles are somewhat like electric cars in that they're primarily suited for close-to-home lifestyles.

Availability of natural gas for vehicles

Natural gas is available in most urban areas, so there is already an infrastructure for the transport of natural gas through pipelines — to homes. Getting natural gas from a gas station is a different matter. Before the majority of gas stations can provide natural gas, money will have to be invested in pumps and tanks. Until then, finding a gas station where you can fill up won't be easy.

Of the more than 150,000 natural gas vehicles in use in the United States, most belong to corporate and government fleets precisely because of the infrastructure challenges. Some cities, for example, have fleets of natural gas vehicles that are serviced at a central hub that provide refilling of the natural gas tanks.

Fortunately, gas stations — or centrally located hubs — aren't the only options. Natural gas is available in homes throughout most of the urban areas in the country. It is possible, with special equipment, to use this as a source to fill up your vehicle's tank. The equipment (called a Phill) is expensive, but some tax credits are available that lowers the cost of the equipment. If Phill is used, fueling your car with natural gas can be very safe and reliable. The biggest benefit? You never have to visit another filling station. That counts for something.

Repairs and safety

Under most conditions, a natural gas vehicle is just as safe as a conventional auto. It might seem more dangerous to drive around with a tank of compressed gas, but conventional autos have a tank of liquid, highly flammable fuel as well. In accidents, natural gas tanks may puncture, but the historical record has not shown that there are more injuries or deaths as a result.

The costs to repair are about the same as for a conventional vehicle, although not all mechanics can fix a natural gas vehicle. In fact, most mechanics want nothing to do with natural gas vehicles simply because they represent a bigger risk if something goes wrong. If you want to own a natural gas vehicle, it's wise to scout out how you'll get it maintained and repaired beforehand. If there's nobody around who will do it, you can find a better alternative.

Here's one more thing to consider: Natural gas stinks like rotten eggs. This may or may not be a problem, but most people find the smell objectionable. You may smell the gas from leaks and also when the tank is being filled up, in which case the smell dissipates quickly. If you continue to smell natural gas, shut the valve off at the tank and call a service technician.

Flex Fuels for Vehicles

Flex fuels are made by mixing *ethanol* (fuel made from corn and other biomass products, as I explain in Chapter 13) with gasoline in varying proportions. The most common mixture ratio is 90/10, or 90 percent gasoline and 10 percent ethanol. E10 is suitable for use in any conventional auto, and auto manufacturers recommend its use because of its high performance, low emission operation. The common term for this gasoline-ethanol mixture is *gasohol.* Most people are not aware that most gasoline supplies now include some ethanol (E10). Half of the United States' supply of gasoline is now blended with ethanol so you've probably been using it without even knowing about it.

Another common flex fuel is E85, which is a mixture of 85 percent ethanol to 15 percent gasoline. This is a much higher ratio than the E10 fuel and implies a different engine tuning setup. For that reason, not all cars can use E85.

The following sections explore the accessibility of ethanol as well as how it's produced and used in vehicles right now.

Sources of ethanol

In the United States, most ethanol is made of corn, but it can also be made of sugar cane and switch grass, among other things. In Brazil, a major producer of sugar cane, for example, sugar cane is the key source for ethanol. In theory, ethanol can be made from most any cellulose, which is a fancy name for plant matter including trees, seaweed, corn, food waste, and so on.

The majority of the ethanol produced in the world comes from corn, which is in abundance in the Midwestern United States. Corn belt states (the Corn Desert) have seen a surge in demand in their main agricultural product due to the intense interest in flex fuel technologies. A lot of federal government subsidies have been directed toward the ethanol industry, and every year there are more and more production facilities coming on line.

Sugar cane is also used for ethanol production, but the quantities produced are so much smaller than those for corn that the contribution is minor. Sugar cane does feature some environmental benefits over corn, so it may be that in the future the use of sugar cane to produce ethanol is a quickly expanding business.

Biomass is a renewable resource. When corn and other biomass products are used to produce ethanol, the crop can be regrown in perpetuity. The production of ethanol, explained in the next section, is a great benefit to farmers, since it results in increasing demand for their products. Yet by using so much of the nation's corn supply to produce ethanol, there is a shortage of corn for the food supply. Corn is a major factor in the world's food supply, and when demand for corn goes up, so do the prices. As a result, there are shortages of food in some parts of the world that can least afford shortages. This is becoming a big political football, since the benefits of mixing ethanol with gasoline (E10 is widespread) are very attractive, yet the fact is, people in the world are starving because of it. The key challenge is determining the right balance between corn used for energy needs and corn used for food.

Producing ethanol

To produce ethanol, soybeans, corn, or other agricultural products are milled down into a fine powder, which becomes a slurry liquid. Enzymes are added to create glucose, and then yeast is added and the material is allowed to ferment (the same process used to make whiskey). This liquid is distilled to separate the ethanol from the base liquid. Because ethanol is produced from biomass, the carbon cycle is closed.

Ethanol is produced by fermenting grains and other biomass materials, so it falls within the category of biofuel (biogas and biodiesel). The production process is similar to the way hard alcohol is made, and there is a vast and mature technology base.

Because ethanol is domestically produced, money and jobs stay in the country rather than going overseas to dubious fossil fuel-producing countries. Ethanol can be produced on site, in small quantities. Farms can produce ethanol out of waste products that would normally not be productive for anything else.

As for the cost (at the pump) of ethanol, subsidies are such a large component of the production cycle that the cost a user pays for a gallon of ethanol is determined by subsidies more than anything else. Ultimately, the government's objective for all alternative-energy sources is to make them competitive with fossil fuels, and so ethanol is more or less competitive. In light of the pollution and political advantages, paying a little more for ethanol-based fuels is worth it to many consumers.

Flex fuel vehicles

Since 1980, the biggest automakers have built over 10 million vehicles that can run on E85 (this is far more than the number of hybrids on the road). In Brazil, a country that adopted a particularly aggressive policy toward the use of flex fuels, over 90 percent of their cars and trucks run off of ethanol/gasoline mixtures in varying proportions. This has significantly reduced Brazil's reliance on foreign fossil fuel suppliers.

Flex fuel vehicles have a major advantage over other types of alternative fuel vehicles in that they operate in basically the same way as conventional vehicles that run on gasoline. The engines used for flex fuel are only slightly different than those for conventional vehicles. When the ethanol mix is low (E10), most conventional engines can burn the gasohol without having to modify the engine at all.

Not all vehicles can run off of a flex fuel, and using it can damage the vehicle's engine and support systems (due to the corrosive nature of the fuel). This is not so much a problem with E10, as with E85. Before you put a flex fuel in your car, make sure it's one that can use it. If you're not sure whether your car is a flex fuel vehicle or not, look for exterior badges and a yellow fuel cap: these mark a flex fuel vehicle. Older versions do not have such markings, and you'll need to consult your owner's manual.

Engine performance and design issues

The driving characteristics of a flex fuel vehicle are the same as for a conventional vehicle. The engine performance of a flex fuel vehicle is the same or better as a gasoline-fueled vehicle. Because flex fuels have a higher octane rating than conventional gasoline, flex fuel cars and trucks experience less knocking (which occurs when the ignition in the engine is not timed just right — the result is an irritating *ping* sound). Flex fuel vehicles may be tuned to optimize the mixture, resulting in even better performance.

Ethanol offers another advantage: It freezes at lower temperatures than conventional gasoline, and so in cold climates it offers a performance advantage when cold startups are the norm.

On the down side, on a gallon to gallon basis, E85 has less potential energy than conventional gasoline. This means that a tank of E85 will propel a vehicle a shorter distance than a conventionally-fueled vehicle. Because flex fuels have less bang per gallon, fuel costs are generally higher on a mile to mile basis. This is partially offset by the improved performance, so it may turn out to be a wash.

Because flex fuels are more corrosive than pure gasoline, special hoses and fittings are required in order to ensure long-term vehicle reliability. Flex fuel vehicles also require special fuel injectors and sensors in order to properly adjust the fuel/air mixture injected into the engine cylinders. These modifications are not expensive, so most existing vehicles can be easily converted to run off of high-mixture flex fuels (higher than E10).

Altogether, hybrids (see Chapter 19) offer more performance advantages than flex fuel vehicles, and will likely outnumber them in the future. But since flex fuels can be used in most existing vehicles, the market for flex fuels will increase as the need to wean ourselves off of foreign oil becomes more imperative.

Accessibility of flex fuels

Flex fuel can be hard to find. While fueling stations commonly offer E10, they don't commonly offer E85. For strictly flex fuel vehicles (that is, those than can run only on E85), it's difficult to find the E85 product. Of the more than 170,000 retail gasoline stations nationwide, only 900 of them sell E85. E85 pumps are also more expensive to install than conventional gasoline pumps. Because flex fuel vehicles can run off of conventional gasoline, most flex fuel vehicle owners are not filling their tanks with flex fuel.

Exhaust emissions

Flex fuel vehicles have lower exhaust emissions than conventional powered vehicles. Ethanol reduces emissions of carbon monoxide, a noxious gas that is imperceptible to the human senses. Since ethanol is made from biomass products, its use does not contribute to global warming.

According to the EPA, the use of ethanol blends can lower carbon emissions by as much as 40 percent, and smog forming nitrous oxides by up to 20 percent. Greenhouse gas reductions of more than 30 percent have been measured. E85, for example, which has a relatively large oxygen content, burns more completely in the combustion cycle, making for better efficiency and lower emissions of both carbon byproducts and other noxious chemicals like sulfur and nitrogen oxides.

Clean Diesel and Biodiesel Vehicles

Diesel is more of a combustion method than an actual fuel (refer to Chapter 5 for an explanation of how diesel engines work). Basically, the process is more efficient than a conventional internal-combustion engine because of the way the fuel is ignited within the cylinder. The process is so simple that a diesel engine can burn a wide range of fuels, and the quality level can be much lower than that of gasoline to achieve high-performance results.

You may have some misconceptions about diesel and its cleanliness as well as how it can be used to power vehicles today. So check out the following sections to get up to date on the technology and advancements diesel is making in the auto industry.

Diesel then and now

Diesel-powered automobiles and trucks have a bad rap. While Europeans have embraced diesel-powered autos due to their reliability and economy, Americans only flirted with them briefly in the early 1980s, during the Arab oil embargo, and then, for the most part, abandoned them. Why? They're smelly and dirty and emit soot and black exhaust (which you can clearly see if you've ever followed behind a car with a diesel engine), leaving the impression that they must be much more polluting than conventional gasoline-driven engines. American manufacturers made matters worse by trying to convert existing gasoline engines into diesel engines and failing on a grand scale. So by the 1990s, diesel-powered cars have pretty much disappeared from American roads, aside from a small number of European imports.

Recent developments in diesel engine technologies have proven that diesels can be much more economical as well as cleaner than conventional engines. While hybrid cars are the current leaders in fuel-efficiency standards, diesels rank a close second. Testing has shown that diesels can reduce fuel consumption by 30 to 60 percent. And these savings do not come at a cost of increased emissions. A study by the National Academy of Sciences showed that diesels could someday supplant gasoline engines as the power source of choice for a majority of autos and trucks.

Clean diesel engines

Historically diesel engines were never designed to reduce emissions or pollution. They were built for power, and they were very good at that. Most large trucks on the road, and most heavy construction machinery, operate with diesel engines. Now that the emphasis on green has grown into an entire industry, diesel engines are being redesigned with new criteria (lower emissions and better performance), and they've been very successful. The following technologies have helped to clean up diesel's act:

- **Common-rail injection system (CRIS):** A breakthrough technology called common-rail injection system (CRIS) made diesels as clean burning as conventional engines. The system creates high-injection pressures that are independent of the engine's speed, and this allows for a much more precise injection, at the exact right time in the engine's cycle. The result is a lot of power with a small amount of exhaust emission. The soot that characterized old diesels is a thing of the past.

- **Turbochargers:** Turbocharging uses the flow of exhaust gases to compress the oxygen being input to the combustion chamber prior to ignition and results in even more efficiency. Catalytic converters have also been added, and a new technology called a *maintenance-free particulate filter* was developed by Mercedes Benz. The result is that diesels, when properly equipped, feature emission performances better than conventional ICEs.

Biodiesel

Biodiesel fuel is a combustible, thick liquid comprised of alkyl esters of fatty acids that come from cooking oil or grease. Being made of biomass, biodiesel is a renewable, domestically-produced energy source. And because it's biomass, the carbon emitted from a biodiesel-powered engine is closed loop: Burning biodiesel doesn't result in an increase of carbon dioxide in the atmosphere.

Biodiesels can be produced from wasted vegetable oils (from restaurants, and even from residences) and other biomass products like soybeans, which are the most common source of biodiesel because they're inexpensive and widely available: One bushel of soybeans can be converted into a gallon of biodiesel.

- ✔ **To make biodiesel from waste products** like used cooking oil and animal fats, the process is similar to the soybean process, but with a couple added steps in the process. Methyl alcohol and sulfur are mixed with the biodiesel raw fuel in a process known as dilute acid esterification. The result is an oil, closely resembling pure vegetable oil, that is further refined into useable biodiesel.

- ✔ **To make biodiesel from soybeans,** the soybeans are processed into an oil, and all impurities are filtered out. Methyl alcohol (among other things) and a catalyst like sodium hydroxide breaks the oil down in a process called esterification. The resulting products, called esters, are refined further into useable biodiesel products.

  Of course, as with corn used for ethanol, there is a political question as to whether the production of biodiesel from soybeans and other widely available food crops exacerbates the world food shortage problem.

The advantages to biodiesel include:

- ✔ Biodiesel reduces dependence on foreign oil.

- ✔ Biodiesel manufacturing plants can be small or large.

- ✔ Biodiesel manufacturing plants can be distributed around the country so that transport to end users is limited, which means that much less fuel is expended simply getting the fuel to the customers.

As for the downside to biodiesel:

- ✔ Unfortunately, the supplies of biodiesel are limited.

- ✔ The idea is relatively new, and production facilities are few and far between.

- ✔ Because production facilities are limited, biodiesel currently costs more than petrol diesel, a situation that will change as more production facilities come on line and efficiencies of scale kick in.

Clean diesel and biodiesel engines

Diesels have been built to operate much cleaner than their older brothers. New diesels no longer emit soot and strong odors. The best news, though, is that biodiesel is a cleaner burning, renewable alternative fuel that is gaining ground every year.

The following sections give you more details about the pros and cons of new diesel and biodiesel engines.

The basics

Biodiesel is commonly blended with petroleum diesel, much like gasohol comes from regular gasoline mixed with ethanol. B50 is an equal proportion of bio-diesel and petrol diesel, while B100 is pure biodiesel. By blending biodiesel with petrol diesel, the same environmental and political benefits are achieved as with ethanol flex fuels: independence from foreign fossil fuel suppliers, a more renewable energy source, domestic production, and so on.

Most existing diesel engines can burn a mixture of biodiesel up to around B20. Newer engines have been designed to burn mixtures up to B100, with minor modifications. The EPA estimates that the use of 100 percent biodiesel will reduce carbon dioxide emissions by 75 percent. A blend of B20, which can be used in most existing biodiesel engines, will reduce emissions by 15 percent.

Consult your owner's manual if you want to burn mixtures higher than B20 because some manufacturers will not honor warranties for engines run higher than that.

Power and performance

By incorporating a much higher compression ratio than a conventional ICE, diesel engines can extract more power from the same amount of raw fuel, and because diesel fuel is typically more energy dense than gasoline, a vehicle can go further on a tank of gas and so refueling is less common. In addition, new diesels are every bit as powerful as their gasoline-fueled brethren. In fact, diesel engines, by their very nature, have more torque, which is useful in accelerating and going up hills. Diesels can also pull heavy trailers without as much strain on the engine.

Biodiesel, on the other hand, doesn't offer the same performance as petrol diesel. What these engines gain in efficiency, they lose in power. They can't deliver the same power levels, and acceleration is slower and hill climbing is more of a chore. On the up side, biodiesel is much safer than an equivalent petroleum product because it's less combustible. The fuel is less flammable than for a conventional vehicle, and it's less likely to explode in an accident. Still, storage, handling, and transport of biodiesel requires a lot of care and attention.

Because diesels are simpler than gasoline-powered engines, they last longer and wear down less (this results in the use of less invested energy in the manufacture of these engines as well — a very attractive side benefit). They also require less maintenance, since they don't need spark plugs and a firing control mechanism. Oil changes are required more often with a diesel because there is more soot deposited on the interior engine surfaces.

Efficiency and emissions

In general, diesel engines are around 20 to 30 percent more efficient than a conventional ICE, with some designs offering up to 50 percent more efficiency.

The combustion of biofuels produces less greenhouse gases than the combustion of pure petrol diesel. Biodiesel also contains less sulfur, responsible for acid rain, especially when made from soybeans and pure vegetable products. Biodiesel also emits less carbon monoxide and other particulate matters. On the downside, diesels emit more nitrous oxides (responsible for smog) than conventional engines, even if biodiesel is used.

And, because biodiesel is biodegradable, it's also not toxic. Spills of biodiesel do not create environmental havoc, although they may be messy and inconvenient.

Chapter 18

Plugging In: Electric Vehicles

Among all the alternative vehicles being developed these days, none attracts more attention than the electric vehicle. People seem to believe that all-electric cars can solve our fossil fuel problems, and many people are emotionally attached to the idea. Electric cars are much simpler than vehicles equipped with conventional, internal-combustion engines (ICE). Electric cars don't need an elaborate transmission, gears, rubber hoses, pumps, and all the other stuff that crowds a modern vehicle's hood, which makes the idea of electric cars very attractive.

But electric cars aren't without problems, the main one being batteries. Simply put, there is currently no viable battery technology that will allow electric vehicles to replace conventional transportation. Whether this will change or not is still up in the air. Batteries have come a long way, to be sure, but they still have a long way to go before all-electric vehicles will be able to provide the performance and range of fossil fuel-powered vehicles.

Nevertheless, all-electric vehicles are being developed at a fast pace. There is a definite market niche right now. In this chapter, I address the technology of electric transportation, the problems associated with batteries, and how and when the current crop of electric cars are best employed.

A Drive Down Memory Lane

The first cars were electric simply because all-electric makes much more sense. Electric vehicles are simpler, at least on paper. The operation is simpler as well: You simply jump in and press the accelerator and you're moving along down the road. Electric cars don't explode and they don't need to be refilled. A dream come true, except for one small problem — or one very big problem — to be more precise: batteries.

Early electric cars, circa 1900

Thomas Edison built an electric car in 1889. His car used nickel-alkaline batteries. Other manufacturers of the time used lead-acid batteries (the type under the hoods of modern cars). Both types of batteries simply fed the battery's electricity directly to an electric motor which powered one or more of the wheels of the car. The next sections discuss the design and disadvantages of the early electric cars as well as the breakthrough in gasoline-powered cars that dealt such a blow to the sales of electric cars.

Early electric car design versus the ICE design

The early electric car designs were simple and very reliable, much more reliable than the ICE-powered cars of the time:

- **Electric cars:** They don't require starting an engine. As soon as electricity is applied to the electric motor, the car goes (hopefully). When more electricity is applied, the car goes faster. You don't need a transmission because electric motors can operate over a wide range of speeds with equal effectiveness. Brakes and steering are the same for electrics and conventional, and so electrics, on net, are much simpler and more reliable. (To find out more about how electric cars work, see the section "Basic Operation of an Electric Vehicle," later in this chapter.)

- **ICEs (internal-combustion engines):** These are much fussier. They require a whole host of support equipment like pumps, hoses, alternators, carburetors, and spark plugs, each of which can break and cause the car to fail. Perhaps most restrictive, ICEs can only run optimally over a limited range of conditions, so they require complex transmissions which not only are prone to failure, but also weigh the car down due to their necessarily massive constructions.

In 1900 a majority of the autos in production were electric. For every one internal-combustion engine vehicle, there were ten electric vehicles. Most of the manufacturers of the day built electric cars and it was believed that the future would be wholly electric. Not a lot of investment money was dedicated to improving gasoline engines simply because electrics make much more sense — except for one big thing: batteries (I know, I repeat myself).

The key problem with electric cars

Range. How far can an electric vehicle travel before it's time to recharge the batteries? That's the big question and the one whose answer doomed the original electric cars and has caused problems for other electric cars ever since.

For these early cars, the driving range was about 40 to 50 miles. To get more range with an electric car, you need more batteries, plain and simple, and that adds more weight, which in turn impacts how far the car can go on a charge. Because the first batteries weren't rechargeable, these first electric cars featured throw away batteries. When the batteries ran out, you simply replaced them with new batteries. Imagine how much that cost! And imagine the disposal problem, except back in those days people didn't really have disposal problems. What you didn't want, you simply tossed into the river. (And we wonder how we got to our current predicament!)

The overall weight of an electric car is around the same for an ICE-powered car despite the weight savings inherent to the simpler design of the electric. The reason is the batteries. They take up a lot of weight, pure and simple.

In addition, batteries can only hold so much charge. As a battery ages, it can hold less and less charge. What this means for an electric vehicle is that, as the vehicle ages, its performance degrades. Add to this the fact that batteries take time to charge, hours of time. Electric vehicle owners must plan ahead at all times.

These are the same problems — range, charging time, weight, and charging capacity — that designers of modern electric vehicles need to combat. Even the most advanced battery technology today can only store a small percentage of the energy potential of a gallon of gas of the same weight and space.

The rise of the gasoline-powered cars

Despite the problems associated with batteries, early drivers preferred the electric cars because ICE-powered cars had an even bigger problem: Getting started.

Starting these early gasoline-powered cars required cranking a handle in the front of the car in order to start it. This was a fussy, often dangerous task because the engine could kick the crank handle backwards and break an arm. (This is similar to the motorcycles that require the driver to jump down on a starter lever in order to get the engine running. Who wants to do that?) In fact, many electric vehicle customers (women in particular) drove electric cars simply to avoid the starting process.

Then in 1912, Alfred Kettering introduced the first self-starting gasoline-powered vehicle, and the difficult task of starting an ICE was no longer in play. After the introduction of the self-starter, electric car sales plummeted and never recovered. Sales of electric cars essentially stopped, completely, and no manufacturers pursued them any longer.

Modern electric vehicles: 1960s to today

In the 1960s electric cars saw renewed investment interest due to the heavy amounts of smog and pollutants that were visible in the skies over cities. Global warming wasn't an issue then, but it was clearly obvious what gasoline engines were doing to air quality. People could see smog, they knew they were breathing it, and they weren't happy about it. They came to the conclusion that there has to be a better way, and electric cars were the obvious option.

But range was still a problem, even with decades of battery development. Cars of this era could go, on average, 100 miles between charges. Some electrical cars were built, but they didn't attract much attention simply because range and performance were so limited. The most logical way to overcome the range problem with batteries was to make the car as light as possible. The lighter the car, the longer the range. But making a car lighter entails eliminating all the unnecessary components and features — what you or I might call the creature comforts, like air conditioners and stereos. As you can imagine, these pared down vehicles were even less popular.

It took federal legislation and regulations enacted by the California Air Resources Board (CARB) to get manufacturers moving into the realm of electric cars once again. In 1990, CARB instituted Zero Emission Vehicle (ZEV) mandates. These mandates required that within just 8 years (by 1998), 2 percent of the vehicles sold in California had to emit virtually zero pollutants. It also mandated that this figure grow to 10 percent by 2003. Many other states enacted similar goals, although California had the toughest standards.

In response, auto manufacturers scrambled to develop cars that could meet the stringent standards, and the race was on. A number of different options were pursued (hybrids, flex fuels, and so on), but it soon became clear that an all-electric vehicle was the best way to achieve the goals. General Motors built the EV-1 and brought it to market with an aggressive advertising campaign. The vehicle worked well, and while it sold modestly, its promise was great enough that a good number were built. Ford, Toyota, and Honda also introduced electric cars.

Then the bottom fell out, for reasons that are open for interpretation (refer to the sidebar "Who killed the electric car, redux" for details of this controversy). Ultimately, the same problem with the earliest electric cars proved the final nail in the coffin for the fledgling electric car effort: Range was no greater than 70 or 80 miles, and then came the long process of recharging batteries.

Electric cars were abandoned by the big manufacturers, and General Motors even refused to renew leases on existing electric cars. The lessees were required to turn their vehicles in, and General Motors destroyed them. Very few electric vehicles are left over from that period, and now they're mostly museum pieces.

The limited interest in electric cars caused CARB to back off its stringent mandates. They decided to address air pollution in different ways, namely requiring gasoline engines to be made cleaner and less polluting.

At this point, if you want an all-electric car, you're pretty much out of luck. Some small companies offer them, but they're dicey and expensive (compared to equivalent gasoline-powered vehicles). However, there are some new designs that are attracting a lot of attention, and there is new development money being invested in electric vehicles. High-end electric cars are being sold that can outperform sports cars made by companies like Porsche and Mercedes. It's interesting to note that one of the early problems was the lack of engine sound. Drivers want their high performance cars to "roar" while accelerating, so in some electric vehicles the "roaring" sound has been artificially added through the stereo system.

Who killed the electric car, redux

There is still a lot of ongoing debate about "who killed the electric car" (the EV-1). Some people claim the electric cars were a phenomenal success story. Others claim that the electrics were abysmal failures because so few customers wanted them. Auto manufacturers need a certain minimal production level in order to make profits, and the number of electric cars being built was nowhere near that level. As to whether they could grow to that level, given enough time, is up for debate.

As with many alternative energy schemes, conspiracy theories abound. Many people believe that the big auto manufacturers never wanted to build electric cars because they would necessarily be smaller, and simpler, and so would result in decreased corporate profits. Conspiracy theorists hold that the electric cars that were sold were intentionally made to be inferior just so the effort would fail, and the auto companies could revert back to building their big, cushy gas guzzlers.

Basic Operation of an Electric Vehicle

The basic operation of an electric car is very simple. An electrical motor is powered by a battery, or bank of batteries. The batteries are generally rechargeable and are very heavy (weight-wise) in comparison to the amount of energy that can be contained within a standard gas tank. The electric motor can be connected directly to the wheels of an electric car. There is no need for a transmission. In fact, a separate electric motor can be connected to each of the four wheels, and this allows for operation even when one of the motors fails, and it also allows for some incredible performance advantages as well. No ICE-powered car can lose its engine and keep on running.

In addition, in an electric car, there is no need for the complex transmission system that an internal-combustion engine requires. Internal-combustion engines work well only when they're running hard, or fast — that is, the crankshaft in the engine is spinning at higher revolutions per minute (RPM). Most of the time — like when the car is going slowly — that kind of power isn't needed. The transmission's purpose is to match the engine's power with the vehicle's speed, which it does with a complex series of gears and shafts. When the vehicle is travelling slow, a low gear ratio ensures that the engine is still operating hard (or fast). As the car's speed increases, the gear ratio changes as needed. (If you've ever driven a manual transmission car, you manually shift from lower to higher gears as the car speeds up. An automatic transmission takes care of these shifts automatically.)

In an electric car, the motor's drive shaft can be connected directly to the wheels of a car, without need for a transmission. This considerably reduces the weight overhead of an electrical car, and it also makes the car simpler and less expensive to maintain.

Not needing a transmission is a distinct advantage that an electric car has over internal-combustion-powered vehicles. In an ICE, as the rotational speed (RPMs, or revolutions per minute) increases, so does the power output (this is why race cars always run near the "red-line" on the tachometer). The faster the engine is going, the more cylinders are firing in a given amount of time and the more power is available (refer to Chapter 16 for an explanation of how internal-combustion engines work).

With electrical motors, the power curve is much flatter than for an ICE. In an electric motor there are no cylinders, and no individual power impulses when each separate spark plug fires. The power output of an ICE is a series of small power blips, but the power output of an electric motor is continuous (okay, let's not get overly technical here — for all practical purposes it's continuous). An electric motor may be operated over a wide range of rotational speeds (RPMs) without sacrificing power output. In addition, you can get the same acceleration from an electric car as a bigger ICE brother. In fact, the acceleration is smoother and doesn't entail the jerking motion conventional vehicles exhibit when the gears are changing.

Batteries

Obviously, batteries are the key component of an electric car: They provide the power that drives the engine. Modern conventional autos feature engines that provide between 60 to over 300 horsepower and more for high performance cars. One horsepower equals around 750 watts, so an electric vehicle is capable of outputting over 75 kilowatts.

Batteries are rated according to their amp hours, or the amount of amps a battery can put out for a specified amount of time. A 1,000 Ah battery can output 1,000 amps for 1 hour (at the batteries voltage), or 1 amp for 1,000 hours. To determine the total amount of power a battery can deliver, multiply the Ah by the voltage and the result is kWhs. *Shelf life* is the amount of time a battery can be stored without connecting it to a load. All batteries will weaken over time, whether used or not. *Maximum deliverable current* is the amount of current a battery can deliver without its voltage declining significantly due to internal resistance. This declines as a battery ages.

Lead-acid cell battery

The most common type of battery in use today is the lead-acid cell. Lead-acid batteries are used in almost every conventional vehicle to start the internal-combustion engine. Once the engine is started, the battery is recharged. These same types of batteries are suitable for driving an electric car, and until much better technologies became available (see the next section on nickel-based batteries), were commonly used.

A plate of lead constitutes the negative electrode and a plate of lead-dioxide provides the positive electrode. Both plates are immersed in a solution of acid, most commonly sulfuric acid. The solution is referred to as the *electrolyte*. When the acid is fresh enough, a voltage difference is generated between the two electrodes, and a load (an electric motor, or a heater or radio) can be connected to receive electrical power. As the battery discharges its power into a load, the electrodes eventually become coated with contaminants and the nature of the acid changes. At some point the process burns itself out, and the battery can no longer deliver power to a load. On the flip side, if a current is delivered to the battery by an external power source (this is commonly called recharging, and is simply the inverse of when the battery delivers its power to a load — in this case, the battery is now the load), the electro-mechanical energy in the battery is restored and the battery will once again be capable of driving a load. The cycle of discharging a battery into a load and then recharging the battery can be repeated many times. At some point, the chemicals become contaminated to the point where the process is simply no longer efficient enough and the battery is discarded or rebuilt.

The amount of power that a lead-acid cell battery can deliver is a function of the acid solution (mass and purity) and the quality of the lead electrode and lead-dioxide electrode, along with their surface area. Over time, the quality of the battery components degrades, and the battery performance decreases.

Nickel-based batteries

Nickel-based battery cells include the nickel-cadmium (NICAD, or NiCd) and the nickel-metal-hydride (NiMH) versions. Nickel-based batteries hold much more charge per unit volume of weight than lead-acid batteries, plus they are capable of more recharging cycles over the course of a lifetime (they're good for literally hundreds of charge/recharge cycles). In some of the newer embodiments, the charge time is decreased, and all of these factors make the nickel-based batteries superior, performance wise, to lead-acid versions.

Nickel-based batteries come in many forms: Cylindrical cells are the tradi-tional battery types that we have all seen so much of. Button cells are the kinds used in cell phones and cameras. Nickel-based cells are also used in space applications, where high integrity packaging is a must. Flooded cells are used in heavy-duty applications, such as electric vehicles.

Nickel-based cells cannot be discharged all the way down to zero or perma-nent damage will result. Therefore, electric cars using these types of batter-ies have controller functions which won't allow complete discharges.

Practical Concerns of Electric Cars

What most people really want to know is how electric cars perform in compari-son to conventional cars. Most people don't really care what's under the hood; they simply want to know what happens when you press the accelerator pedal.

One of the biggest changes you'll notice as a driver of an electric car is that you can't just say to yourself, "I'll charge in the morning." You need to plan ahead because, unlike internal-combustion vehicles, electric cars cannot simply be filled up at the corner gas station. It takes time to charge the batteries, and so foresight is a necessary component of electric car ownership. The following sections outline some of the other practical considerations of owning an electric car.

Performance

All-electric cars can be made every bit as powerful as a conventional auto, and some all-electrics are even more powerful. The problem with power is that it takes a lot of juice out of the batteries in a very short time, and recharging becomes necessary. The ride of an electric car is also smoother and quieter than the ride of a gasoline-powered car.

Keep in mind that batteries are temperature sensitive and lose some of their output power when temperatures fall below freezing. The more features you use while you're driving your electric car, the shorter the range as well. While this is true for conventional autos, it takes a lot less time to refill a gas tank than to recharge, so the burden is less.

All-electric cars are particularly good for some applications: namely, city driving and driving short distances. This is one of the reasons why these vehicles are most common in cities, where commute distances are shorter than in the suburbs and where the narrow city streets and the crush of traffic make the generally smaller all-electric vehicles easier to maneuver. But there are a few things that electric cars aren't so good at:

✔ **Hauling loads:** Don't expect to see electric trucks anytime soon.

✔ **Including creature comforts:** Electric cars are necessarily spare and simple. They will not come with a lot of options and accessories simply because these limit the range of the vehicle.

✔ **Long distance travel is generally out of the question:** Why? Because there are no charging stations along the roadways — a situation that may change in the coming years. One scheme attracting interest is a battery-swap arrangement where you simply hand in your exhausted batteries and get freshly charged units installed in their place. This kind of swap will enable a quick "fill-up."

The performance of an electric car is really based on the batteries it uses. Because battery technology is constantly improving there will likely be much better batteries in the next few years. It's likely that you'll be able to achieve a much better performance in a couple years by simply swapping up to a higher grade of battery when your old batteries need replacing. No other vehicles on the road can offer this promise.

Service and maintenance

Electric vehicles are extremely simple compared to any other type of vehicle, particularly ICE-powered vehicles. Service is easy, and straightforward. There are no tune-ups needed, no radiator flushes, no change of spark plugs, no filters. A single person can remove and replace an electric motor, and there are very few support systems needed (like carburetors, radiators, and so on) that need care.

Because electric cars are rare, finding qualified service technicians can be prohibitive. This will change over time, but it's going to take many years before the infrastructure is in place to adequately handle electric cars.

Battery charging and replacement

Keep in mind the following considerations about charging and replacing the batteries in electric cars:

- ✔ **The batteries in electric cars need to be charged regularly.** Most of the new electric vehicles being developed include an on-board charging scheme so that all you have to do is plug your electric car into a wall outlet that provides utility power.

- ✔ **If you plug your electric car into an outlet in your garage your electricity bill is going to go through the roof.** If you're on a tiered rate structure (the more electricity you use, the higher the cost) you could be in for a major shock when you see your bill. If you can sign up for time-of-use (TOU) metering, an electric vehicle will work better, if you charge it only at night.

- ✔ **Even with the new-and-improved batteries, batteries on electric cars will eventually wear out and need to be replaced.** The best batteries will last over five years, with some auto manufacturers warranting batteries for the lifetime of the car. Electric cars use a good number of batteries, so changing to new ones can be very expensive. It's not a difficult job to change batteries, so the labor costs are low, and most hybrid and electric car manufacturers will not warrant a battery pack unless it's changed by an approved service pro.

- ✔ **Because batteries contain a lot of harmful chemicals you can't just throw them away.** They must be disposed of properly (at approved disposal sites they will take batteries, but they charge for this service because they have to dismantle the batteries and dispose of the chemicals via specialized processes).

Environmental impact

Electric cars produce no tailpipe emissions and only miniscule amounts of evaporative emissions (emissions from evaporating fluids, like gasoline and oil in conventional autos).

Despite the fact that electric cars need to be charged via grid power, which uses coal and other fossil fuel power sources (see the next section "Where you get your charge" for info you need to know about that), mile for mile, the pollution effect is much less for electric cars because they are smaller and more efficient, and widespread use of electric cars will decrease the aggregate pollution levels of transportation.

Where you get your charge

The electricity to recharge the electric car's batteries has to come from somewhere, and that somewhere is generally the electrical grid. Grid power is mostly derived from the burning of coal (52 percent, in the U.S.), a situation that's not likely to change over the next few decades for reasons that I explain in Chapter 4. Coal is widely available, and the supplies are mostly domestic.

Coal-fired electrical power is one of the most polluting sources. Recharging an electrical car from the grid is not necessarily going to solve our fossil fuel problems. In fact, it might make them worse if enough people buy electric cars because the grid will come under heavier and heavier use. The only way that electric cars will mitigate our dependence on fossil fuels is if they're so efficient in their battery-to-wheels operation as to offset all of the inefficiencies encountered along the energy line. This means that electric cars need to be small and spare, with few creature comforts.

Of course, electric vehicles can be charged by sources other than the coal-fired grid. If your utility derives its electrical power from a nuclear reactor, the net pollution created by an electric car can be very low. Some other possible scenarios include

- ✔ **Using large solar arrays mounted over parking lots.** The cars beneath the arrays can be charged during the day when the cars are parked beneath the panels. This scheme reduces pollution levels even further. It has another advantage as well: Cars parked beneath the solar arrays are shielded from the sun, and are much cooler to enter after a long day of work.

- ✔ **Using wind power:** In the same way that solar power can recharge batteries, wind power can be connected to the batteries for recharge. Wind power has the added advantage of being available at night, or in bad weather.

The problem is that these sources of electrical energy are not reliable: You're in trouble if you can't charge your car during the day when the sun's out, or if the day is calm or cloudy.

The chemicals in batteries

Lead-acid batteries and NICAD batteries contain a lot of nasty chemicals. The sulfuric acid in a lead-acid battery causes noxious fumes like sulfur dioxide. Hydrogen gases are also produced, and these are explosive when exposed to sparks or flames. Lead and cadmium are heavy metals and cause environmental harm when they enter the water system. NiMH batteries are less harmful than NICAD batteries and because of this there is a push to eliminate NICAD altogether.

Efficiencies

Batteries are inefficient. It takes around 10 to 20 percent more energy to charge a battery than the energy that can be derived from that battery. A 12-Volt battery (the most common voltage output available — the type used in conventional autos) rated at 2,000 Ah required more than 24,000 Whs, or 24 kWhs, to fully charge. This amount of energy will drive an electric car for around 50 miles. The average cost of a kWh of energy from the grid is around 10 cents, so the cost of driving an electric car 50 miles is around $2.40. This varies widely, however, and doesn't take into account the cost of new batteries.

Electric car safety

Rumor has it that electric cars are more dangerous than conventional autos, particularly in accidents. But this hasn't been borne out by the record.

Size issues

Electric cars are generally smaller than their conventionally-powered bigger cousins. In fact, tiny cars are ideal for electric power. There are a number of small, one or two passenger vehicles on the market, which can transport one or two passengers over short distances.

Because electric cars are necessarily small and spare, safety is an issue. When a small vehicle gets into a crash with a larger vehicle, the large vehicle will generally win. There are so many large-sized, ICE-powered vehicles on the road now that it will take decades before the average size of a vehicle on American roads decreases. Even if the vast majority of vehicles sold in the next few years are small electrics, there will be a lingering size disparity.

The insurance industry weighs in

Insurance rates for electric vehicles will be higher than for conventional autos for two reasons. First, the increased hazard of higher voltages will result in more injuries. Second, electric cars are rare, and they will be for the foreseeable future. Insurance companies don't like rare cars because they are harder to replace. In addition, the risk that electric vehicles will once again disappear into the vapors of time means that insurance companies may not even be able to replace your electric vehicle, even if they want to. This puts the venture at risk.

Voltages: The shocking truth!

Electric cars also use higher voltages to run. But higher voltages are more dangerous because shocks are easier to get. You can get quite a jolt messing around under the hood of an electric vehicle, and who can resist messing around under the hood of a new car?

Manufacturers have incorporated a good number of safety features into the designs of their electric cars. Batteries are well sealed and the high voltage lines are clearly marked, and when possible, insulated to the point where getting a shock requires one to really try to get a shock.

Economics of electric cars

Electric cars today are very expensive. Production is limited to small manufacturers who don't achieve economies of scale. This may change as more electric cars hit the road, but right now you'll pay an arm and a leg for an electric vehicle. Tax breaks, subsidies, and rebates are available from different government agencies, and this helps to drive the cost of driving an electric car down even further.

The cost of driving an electric vehicle, mile per mile, is less than the cost of driving an internal-combustion vehicle. In some cases, the costs are nearly one tenth. Despite the inefficiencies of the electric grid in providing power to recharge batteries (refer to the earlier section "Where you get your charge"), electric motors are very efficient compared to ICEs, and so even in regions where electrical utility rates are high, driving an electric car is less expensive than driving an ICE-powered vehicle—at least on a mile per mile basis.

And last but not least, the ultimate determining factor of the cost of an electric vehicle is the cost of batteries. There is no way to tell if battery costs are going to rise or fall.

There is an inherent risk in buying an electric car. The same fate may be in store for electric cars as was experienced in the 1990s. New models will be introduced but a few years later the manufacturers may decide to abandon electric cars once again. In particular, hybrids, discussed in Chapter 19, show more promise because they neatly and effectively solve the range problem of electric vehicles. If you buy an electric car, you may be left out in the cold with a vehicle that nobody can service, and nobody wants to take off your hands.

Chapter 19

Hybrid-Electric Vehicles

· ·

In This Chapter

▶ Checking out hybrid history

▶ Finding out more about hybrid features

▶ Taking a look at parallel and series hybrid technologies

▶ Driving a hybrid

· ·

All-electric vehicles are severely limited by the fact that batteries can only contain a certain amount of energy, and when that is used up the batteries need to be recharged, which takes time (hours) and special facilities (you need to either plug them in, or change batteries). Internal-combustion vehicles are limited in that the efficiencies of internal-combustion engines (ICEs) will never be as good as electric drive, and the emissions will never be as low as electric vehicles. Hybrids — cars that combine electrical drive with an internal-combustion engine — achieve a harmonious balance between the best of both worlds.

Of all the alternative transportation options available today, hybrids promise to become the most widely accepted due to their optimal balance between the best of internal-combustion-driven vehicles, and all-electric vehicles. In this chapter I describe the various types of hybrids, and their pros and cons.

History of Hybrids

Hybrids are nothing new. Some of the first cars were hybrids because early designers recognized that gasoline engines have certain characteristics that are nicely offset by electrical power. As early as 1900, hybrid cars were on display at the Paris Auto Salon, a huge international show that featured the most modern auto technologies. Commercial trucks used hybrid technologies as early as 1910, with mixed results.

Witnessing the fall of hybrid hoopla

The hybrid concept started to die out (as did the early electric cars; refer to Chapter 18) when the self-starting motor was invented for use on internal-combustion engines. Historically, one had to go to the front of an internal-combustion auto and crank a handle to start the engine, a dangerous task that often caused injuries and one that was nearly impossible for a woman to perform. The self-starting motor was itself electric and made it possible to start an internal-combustion engine with the simple twist of a key. From that point on, internal-combustion engines were widely accepted and ICE-driven cars were on their way to the universal acceptance they now enjoy.

As the cost of gasoline started to come down with improved drilling and refining techniques, hybrids lost even more favor because they were overly complicated, with dual power sources and all the ancillary equipment required to support the dual power sources.

In a way, an electric starter motor is not much different from a hybrid drive motor because the starter motor required a battery bank that needed to be charged in order to drive the electric motor. So it might be said that conventional autos are hybrids, although they don't derive drive power from the electric starter motor.

Watching hybrid research pick up

Fast forward a few decades from the early 1900s when the excitement over hybrids was overshadowed by the self-starting motor to a time when society was becoming more environmentally aware. With increased attention to the high pollution levels of internal-combustion engines, the major auto companies began experimenting once again with hybrids in the 1960s. Research and development money was scarce, however, and the technology was little more than a curiosity exhibited at auto shows. Hybrids were touted as the cars of the future, and the future kept getting pushed further and further out. But then things started picking up:

- **1970s:** In the mid 1970s, the U.S. Energy Research and Development Administration initiated a program to develop hybrid technologies. Congress enacted the Electric and Hybrid Vehicle Research, Development and Demonstration Act.

- **1980s:** A German manufacturer unveiled a workable hybrid with a 13 horsepower engine driving the rear wheels while a 123 horsepower electric motor powered the front wheels.

✔ **1990s:** The Clinton administration created a federal initiative called the Partnership for a New Generation of Vehicles (PNGV). A consortium of industry and government agencies set about to develop an extremely low pollution auto with a goal of over 80 miles per gallon. Different concepts were built and tested, but it became clear that the only way to achieve the goal, while still accomplishing a workable performance vehicle, was through hybrid technology. A few working prototypes were built, but none of them went into production.

It didn't help that the research and development money was limited to American manufacturers, and it's ironic that now the Japanese car companies are ahead of the Americans in terms of hybrid technologies. A great opportunity was squandered, which is unfortunately a typical conclusion in the energy industry.

The first modern hybrids

By the end of the 20th century, Honda had beaten Toyota in introducing the first economically viable hybrid to the American market with its Honda Insight (1999). The car offered incredible Environmental Protection Agency (EPA) performance ratings for both pollution levels and miles per gallon. Toyota quickly followed up with their Prius, which is the best selling hybrid on the road today. Both Toyota and Honda developed commercially viable autos while the American car companies held firmly to their ICE-powered mainstays.

Gaining popularity

Early acceptance of hybrids was slow to take root. Their performances were very good. They could accelerate and outperform similar ICE-powered vehicles, but the prices were (and still are) higher due to the fact that they use dual power systems. Toyota and Honda did not have the production capacity in place to ensure lower prices due to manufacturing scales, and neither company promoted their hybrids with much vigor. But most of all, when hybrids first hit the market, gasoline prices were still very low (compared to today's prices). There simply wasn't enough demand for hybrids, with their increased gas mileages, and they were little more than curiosities driven by "tree huggers."

But when the world started paying more attention to global warming and pollution, interest in hybrids began to gel. The media attached themselves to hybrids, and even if Toyota and Honda didn't promote their technologies they got a lot of free publicity (who can shun free publicity, especially when it's favorable?).

The real push for hybrids came when it became clear that all-electric vehicles were simply not going to work for most auto applications (due to their limited ranges). Plus, electric vehicles are still limited by battery technology problems. Go to Chapter 18 for the details on the triumphs and travails of electric cars.

Two Key Hybrid Features

Two features mark all hybrids: regenerative braking and the use of more than one power source, the most common being a combination of a battery and internal-combustion engine (ICE). The next sections explore the facts surrounding these two key features of all hybrids.

The onboard ICE and battery combo

Any vehicle that uses more than one power source is classified as a hybrid, but the most common version of the technology is for a vehicle that combines electrical drive with a heat engine (ICE, or fossil fuel combustion engine, such as those addressed in Chapter 5). The onboard ICEs used in modern hybrids are designed for optimum efficiency to reduce fossil fuel consumption and emissions to the lowest possible levels. With a hybrid combination of electrical and ICE drive, it's possible to run the ICE engine at a very restricted RPM range, and this further enhances the efficiency of the vehicle because the engine can be run at its most fuel efficient, low emission level.

The ICEs used in hybrids are simpler than the ones used in gasoline-powered vehicles because they don't need to run over large operating ranges. Simple means less expensive. The internal-combustion engines used in hybrids are unsuitable for use in purely ICE-powered vehicles. A controller determines how much ICE power and electrical power are needed at any given time in order to increase efficiency to the highest level possible. When the battery is sufficient to power the vehicle, hybrids run under battery power. The heat engine (ICE) is only run when the batteries need recharging or when the vehicle needs more power to crest a hill or pull a heavy load. As a consequence, in hybrids, the ICE engine is constantly started and stopped as conditions warrant.

Obviously one of the design goals is to use the smallest, lightest weight batteries possible, but this then means that the ICE comes on more often, for longer periods. So there is a critical balance between the size of the electric power plants (the battery) versus the fossil fuel power plants (the onboard ICE). It is this balance between battery size and capacity, heat engine size (with the attendant fuel tank) and control algorithms that determines the success of the hybrid vehicle. At this point, there are many different combinations of battery size, ICE engines, and controllers. Over time, there will likely be a convergence to a particular design that works best under the widest possible conditions.

And as with all electric vehicles, improving battery technologies will allow for better performance and less expensive up-front costs.

Regenerative braking

Regenerative braking is a standard feature on all hybrids, and increases their efficiencies in relation to conventional vehicles.

When a car is going down a hill, gravity provides the drive power. There is no load on the engine, whether it's an electric motor or the internal-combustion engine. In fact, when a car goes down a hill or experiences braking, a lot of available energy is dissipated in the brakes and turns into wasted heat.

A regenerative braking system captures that wasted kinetic energy and converts it into an electrical signal that charges the battery bank, thereby making the overall vehicle operation much more efficient. The wheels are connected to the electrical generator instead of the brakes, and the generator's "back pressure" causes the vehicle to slow down. This not only recaptures energy that would otherwise be wasted in the brake drums and pads, but it significantly extends the life of the brake pads as well (you can't eliminate brake pads altogether because in hard stops the regenerative system simply can't stop the vehicle fast enough).

Hybrid Technologies: Series and Parallel

There are two basic versions of hybrid electric vehicle (HEVs): parallel designs and series designs. In *series designs,* the vehicle is basically all electric, with the onboard ICE serving only to charge the battery bank. *Parallel designs* use complex transmissions that can channel either the electric motor power, or the ICE power, directly to the wheels.

With both types of hybrids, the raw fuel options are also much greater than for conventional ICE vehicles, and the pollution mitigation potentials are greater as a result. Renewable fuels are more commonly used in HEVs than in conventional vehicles. Engines can be powered by gasoline or methane, propane, E85, petro-diesel, or hydrogen (see Chapter 17 for info on flex fuels and Chapter 15 for details on hydrogen fuel cells).

Series design hybrid vehicles

In the series design hybrid, all the motive force is derived from the electric motors. The primary engine (ICE) serves only to charge the battery bank, which drives the electric motors. Electric motors are limited in terms of the amount of torque and power that they can deliver, and so the performance of series design hybrids is limited. They are, however, much simpler than the parallel designs, explained in the next section, and therefore less expensive and easier to maintain.

In a series HEV, the motive power is obtained via electric drive motors driven from a battery bank, or several battery banks that use different types of batteries. This is exactly like an electric vehicle (EV), but instead of relying solely on recharging of the batteries through an external source (like the grid) a small internal-combustion engine drives an electrical generator that recharges the battery bank. This means that the car can be driven much greater distances since the primary fuel is gasoline which can be obtained at any gas station.

The gasoline engine in a series HEV runs at a constant speed, which is far more efficient than the variable speeds that are required from ICE-driven vehicles or parallel design hybrids. The pollution levels are less, and the mileage efficiency is greater.

Parallel design hybrids

Parallel designs are far more common because they can deliver more power to the wheels, and drivers accustomed to ICE-propelled vehicles demand acceleration and handling. In a parallel hybrid, both the electric motors and the internal-combustion engine drive the powertrain. The power burden automatically shifts between the two power sources as the road conditions and load conditions change.

At low speeds, when relatively little power is required to propel the vehicle, the electric motor is used. At high speeds, when a lot of power is needed to overcome frictional forces, the internal-combustion engine is used in conjunction with the electric motor. A processor determines the mix of power sources that are most efficient for the load conditions.

The parallel design can be used to power heavy vehicles such as buses and trucks, thereby making the hybrid much more versatile than an all-electric vehicle.

Driving and Owning a Hybrid

People like their cars, and they like the way they drive and handle. A big question is how the performance of an HEV measures up to conventional autos. In this section I describe the important aspects of HEV day to day operating characteristics. In some respects HEVs are actually better and more fun to drive. There are, however, some drawbacks worth noting.

Performance and handling

People expect that hybrids will perform much differently than conventional autos, but this is not the case. In fact, under most driving conditions, hybrids can actually outperform conventional autos. Some of the most expensive auto makers are now offering hybrid versions of their conventional standbys, and there is no performance sacrifice. Lexus, for instance, is now offering hybrid versions of their top-of-the-line luxury sedans, and they would not be doing this if they weren't confident that buyers would react positively. The Cadillac Escalade is fast and powerful, and General Motors is advertising the fact that the Escalade gets better gas mileage than a Subaru (the heretofore energy-efficient brand name).

Electric cars are limited in range, so it's no surprise that, because of their reliance on internal-combustion engines to recharge the battery "on the go," hybrids can travel much farther than electric cars. What many people may find surprising is that they can travel significantly farther than ICE-powered vehicles, too. The typical conventional auto needs to be refueled every 300 to 400 miles, but hybrids routinely offer ranges greater than 500 miles.

Other things to know about the performance and handling of hybrids include the following:

- ✔ The hybrid's ICE is sized to transport an average, not a peak, load. While this affords a large measure of efficiency to the overall operation, it also means that hybrids aren't made for speed and power. To get the most fuel efficiency and range out of a hybrid, however, requires modifying your driving habits to maximize the amount of time the hybrid's electric engine is providing the power. For many, that means changing what they expect out of their car: sheer speed and power or increased mileage and decreased emissions.

- ✔ Most of the gains in pollution reduction and mileage increase come when the hybrid is operating in all-electric mode, and so hybrids are more advantageous in city driving (short trips) than rural driving. This is the opposite of conventional autos which offer better performance in highway driving than city.

✔ People want to feel the power buildup as they accelerate. To the drivers who are accustomed to the "normal" engine sounds — the roar of the engine when the car runs fast, the sound the engine makes as the transmission moves through the necessary gears, and the low rumble of an idling — this seemingly random starting and stopping of the ICE can be a little off-putting. It doesn't offer the same sound feedback that most drivers have come to expect (and a few actually love).

Maintenance

Hybrids use ICEs, just like conventional autos, and these need to be kept tuned and running at peak performance. And hybrids require the usual routine of oil changes, wheel alignments, and so on. But they also require periodic electrical testing to make sure the control systems are running the way they should. Above all else, the large battery packs hybrids use need to be maintained and tested on a periodic basis. Not only are they filled with dangerous chemicals, but they wear out, especially if they're abused. They're also very expensive to replace. And in extremely cold weather, battery banks lose power and may not operate properly.

Hybrids are more complicated than conventional autos, there's no denying it. Unlike a problem in a conventional auto, which can be reliably determined by any competent mechanic at an auto repair shop, the cause of a problem in a hybrid isn't very intuitive, and there are relatively few technicians trained in hybrid technologies. Electrical systems are difficult not only to comprehend, but to repair and troubleshoot. For these reasons, the cost to repair a hybrid may be very high. And if you're out in the middle of the boondocks, you may not even be able to find somebody who can repair your broken down car.

There is very little infrastructure dedicated to the parts and servicing of hybrids, at this writing. This will change over time as hybrids become more common. Even if the repair infrastructure improves dramatically, hybrids will still be expensive to maintain. Parts may be difficult to get, unlike with conventional autos where parts stores are on every corner, and the workers behind the counter know precisely what you need and how to install it.

Another maintenance consideration is how to dispose of spent batteries. Disposal of batteries is expensive. You can't just throw them away; they need to be specially processed by qualified recycling centers and these may be far and few between. While the shops that swap battery packs will dispose of the batteries, it's expensive and ultimately the consumer is going to pay the price.

Environmental impacts

A key fuel-saving advantage of a hybrid is the automatic shutoff and start procedure for the gasoline engine. When the vehicle is stopped, the gasoline engine is off, which saves a lot of gasoline. It is estimated that if every ICE engine in the country were simply turned off when the car was not moving, aggregate fuel consumption would go down 8 percent. Of course, turning off your engine every time you stop isn't practical with a conventional vehicle, but with the hybrid, this very thing happens automatically. In driving short distances or creeping along in stop and go traffic (a common scenario of city driving), the ICE engine never even comes on at all.

This, in addition to the regenerative braking capability and smaller, fuel efficient internal-combustion engine, allows hybrids to achieve significantly better gas mileage figures (fuel economy is often 35 percent or more better for a hybrid than a conventional auto of the same size), along with significantly reduced pollution levels.

The first hybrids cut vehicle emissions of global warming gases by a third to a half, in comparison to ICE-driven vehicles of the same size and class. More recent models have cut emissions by over two thirds, and technological advancements promise even better performances. Bottom line: The overall pollution generated by HEVs of both types is much less than for a conventionally-powered vehicle using an internal-combustion engine, even when the electrical pollution from the grid (if it is used to charge the battery bank) and from the on-board batteries is taken into account. Hybrids emit only 50 percent of the carbon dioxide of conventional autos, and only 10 percent of the carbon monoxide.

The economics of hybrids

HEVs cost more than conventional vehicles. The up-front investment is higher than for an equivalent internal-combustion-driven vehicle, and it takes some time to recoup this via better fuel economy. Despite the higher cost of hybrid vehicles themselves and the higher maintenance costs, the cost of driving a hybrid vehicle is much less than for a conventional vehicle because hybrids are smaller and more efficient. In addition, tax breaks, subsidies, and incentives are available from many different government agencies, thereby making the economics even more attractive.

If carbon taxes are instituted, the price differentials between hybrids and conventional autos will disappear, and hybrids may someday outnumber conventional vehicles on the road.

Buggaboos and boogeymen: Unwarranted fears

There are a number of negative myths that seem to follow hybrids, but most of them are unfounded. The track record is still being tallied, but hybrids have proven themselves safe and economical over the last decade. Here are a couple of the most prevalent myths.

They're dangerous

There is a recurring myth about hybrids that simply won't go away. The story holds that police and emergency personnel are put at undue risk from hybrids because their electrical systems operate at higher voltages. But the threat hasn't panned out.

Hybrid battery packs are well sealed and can take a collision without rupturing. The high voltage systems are well contained and marked so that emergency personnel know what they're confronting. With training, emergency personnel know exactly how to deal with a hybrid accident. They know to remove the ignition key, which disables the electrical system.

As with all cars sold in the United States, hybrids must meet the stringent requirements of the federal motor vehicle safety requirements. The same seatbelts, airbags, and electronics are used on hybrids as with conventional vehicles. In addition, hybrids have more intelligence than conventional autos and there are additional safety features built in that take advantage of the increased intelligence.

They're too new to take a chance on

One of the biggest hurdles right now for the hybrid industry is the notion that the technology is going to mature quite a bit in the next few years. We have all seen this effect with computers and large screen TVs, and we feel as though the same effect will apply to hybrids. There is a maxim that says, "never buy the first generation of any product." This is true of cars as well as electronics.

A marketing poll found that people are leery of hybrid cars because of their relative complexity. But the fact is, hybrids have been around for a long time, and the technologies are well proven.

Chapter 20

Hydrogen and Fuel Cell Vehicles

. . . .

In This Chapter

▶ Exploring the promise and challenges of hydrogen as a fuel source

▶ Checking out vehicles that use pure hydrogen and those that use fuel cells

. . . .

*W*hen a majority of experts look into the crystal balls, they see two viable options for solving the world's fossil fuel problems:

✔ Use the same fuels that are used now — gasoline and diesel from petroleum products — but supplanted by ethanol, biodiesel, and other renewable energy sources, particularly those derived from biomass. Hybrids of all stripes fit into this category (refer to Chapters 18 and 19), as does diesel in all its various forms (see Chapter 17).

✔ Use hydrogen. Hydrogen fuel cells (covered in Chapter 15) convert chemical energy directly into electrical energy and leave no trail behind other than a small amount of water and some carbon dioxide — much less carbon dioxide, in fact, than conventional internal-combustion engines (ICE) and clean diesel. There are no particulates, nitrous oxides, carbon monoxide, or any other environmentally damaging emissions. Most experts believe the future lies in this scenario.

The big automakers have been toying with fuel cells for use in transport since the 1960s, when fuel cells were used successfully in the space programs. But only in the last 15 years have major inroads been achieved. These were brought about by the zero-emission vehicles (ZEV) mandates passed by California. Today, the race is on to develop an economical hydrogen-powered vehicle design that can be sold commercially.

Although I describe hydrogen fuel cells in Chapter 15, I get down to the nitty-gritty details in this chapter of how fuel cells can be used specifically in vehicles.

Fuel Cell Fundamentals and Challenges

Hydrogen fuel cells operate similarly to a common battery. There are two electrodes, a positive and a negative. Oxygen passes over one of the electrodes and hydrogen passes over the other. A complex chemical reaction occurs and produces electricity, along with heat and water. The heat may be used to provide other vehicle functions like internal space heating and maintaining temperatures in the control system (electronics like to operate at specific temperatures in order to achieve optimum efficiencies). The water may simply be discarded, without environmental damage. Unlike batteries that need to be recharged, however, hydrogen fuel cells will run as long as oxygen and hydrogen flow over the electrodes.

Two benefits stand out when you're talking about fuel cells:

- Hydrogen, when combusted with pure oxygen, is entirely pollution free; the only byproducts are water and heat.

- Hydrogen and oxygen are everywhere.

Sounds like a dream come true. And it is — if a few wrinkles can be worked out. The following sections outline those wrinkles as well as the basics and benefits behind fuel cells.

Dealing with hydrogen

The "fuel" needed for fuel cells is hydrogen and oxygen, and fuel cells simply convert the molecules in these elements directly into clean, powerful electrons (electricity). In the next sections, I describe the challenge in obtaining the "fuel" needed for fuel cells as well as a promising method for extracting and storing hydrogen, the more fickle of the two necessary elements.

Hunting and gathering

Gathering oxygen, obviously, is a no-brainer. It's everywhere and in abundant supply. Hydrogen, on the other hand, while abundant, is a little more problematic. It can be a very dangerous substance because it reacts so strongly with other elements. That's why hydrogen in its pure form is very rare in nature.

The whole trick to making hydrogen fuel cell vehicles economically viable is in developing effective means of finding, producing, and storing hydrogen on a scale which can be used in vehicles. In Chapter 15, I describe some of the ways that hydrogen can be produced, like steam reforming, plasma waste, electrolysis, and so on. The common goal of all these methods, very basically, is to use heat, pressure, or some type of chemical reaction to separate, or extract, the hydrogen from the other elements that it naturally combines with.

Some methods show more promise than others, but suffice to say that none of the methods has yet proven practical because the cost of the processes is so prohibitive. But with the increasing attention brought to bear on hydrogen fuel cells, research money is being invested that will not only bring the costs down, but will isolate and identify the most viable long-term methods of hydrogen extraction.

Reforming

Perhaps the most viable alternative to date is to simply use our existing fossil fuel supplies for the production of hydrogen via the process of *reforming*. This technology is well understood, albeit complex and expensive. Reformers basically *crack* the raw fossil fuels down into constituent elements. Cracking is a process where complex organic molecules (fossil fuels as well as carbohydrates) are transformed into simpler molecules.

The corner gas station could be converted into an on-site hydrogen production facility, using the raw fuels that they now dispense. Better still would be to have a reformer directly on the vehicle so that hydrogen would always be available (with an occasional fill-up at the gas station of a suitable raw fuel used by the reformer).

If reformers can't be developed (and therefore hydrogen can't be created on site), huge investments in infrastructure will be required to get the hydrogen to filling stations.

The high cost

Like any nascent technology, hydrogen technologies are expensive, making the production, operation, and maintenance of hydrogen-powered vehicles expensive as well. Adding to the cost is the fact that hydrogen itself is an expensive fuel.

The economics of hydrogen-fueled vehicles are prohibitive not only because the raw fuel costs are high, but also because hydrogen engines are expensive. The upfront costs of a hydrogen-fueled vehicle are very high. Even though tax breaks, subsidies, and other incentives are available for those hearty souls who wish to be on the leading edge of technological innovation, one must be very keen on achieving the zero-pollution performance in order to shell out this kind of money.

Despite the technical hurdles, hydrogen cars are going to hit the market soon. The cost will be high, and the operating costs will be even higher. But enough people are attracted to the notion of emitting nothing more than water from their cars that markets will develop. And as the markets develop, prices will inevitably come down. At some point, hydrogen fuel cell vehicles will be price competitive with conventional ICE-driven vehicles and then the entire world will be driven by hydrogen. . . . Maybe.

Benefits of hydrogen and fuel cells

Despite the issues of cost and hydrogen extraction, fuel cells offer a number of benefits:

- ✔ Hydrogen engines are extremely efficient compared to fossil fuel alternatives. Although hydrogen fueling stations are far and few between today, once hydrogen is widely available, this option promises excellent performance and pollution advantages.

- ✔ The existing pipeline infrastructure for natural gas could be converted to use for hydrogen transport to service stations. This would help considerably with the economics.

- ✔ Hydrogen can be produced on site. It's possible to produce hydrogen at home, using electrolysis or other methods. In fact, hydrogen can be produced nearly anywhere, even on the vehicle itself. Some scientists believe that hydrogen-fueled vehicles will become the main source of transport when on-board hydrogen producing technologies mature enough to become economically viable.

Hydrogen-Powered Vehicles

Despite its promise, the current hydrogen technology used in vehicles is still very immature. Today, two options are being explored:

- ✔ **Using hydrogen to power internal-combustion vehicles.** Called *hydrogen fuel vehicles* (HFVs), these cars don't use fuel cells; they run on pure hydrogen.

- ✔ **Using hydrogen fuel cells in electric vehicles.** These cars, called *fuel cell vehicles* (FCVs), use fuel cells in addition to or in place of batteries in electric vehicles.

The following sections go into more detail on these vehicles, addressing not only how these vehicles work, but also how designers have addressed the challenges posed by relying on the volatile element of hydrogen.

Hydrogen fuel vehicles

Conventional fossil fuel vehicles can be modified to run on hydrogen in the same way that they can be adapted to run on methane or propane. Hydrogen fuel vehicles (HFVs) do not use fuel cells; they run on pure hydrogen in a combustion process. BMW is one of several automakers currently developing production cars that use hydrogen in an ICE. The engines can use either hydrogen or fossil fuels because a strictly hydrogen-powered car would be completely impractical due to the unavailability of hydrogen.

The BMW hydrogen/fossil fuel prototype drives exactly like its purely fossil fuel cousin; there is no degradation of performance. The car accelerates from zero to 60 mph in 9.5 seconds, which is not rip roaring, but it's typical for a big sedan. Running on hydrogen only, the 12-cylinder engine delivers 260 horsepower. It has a range of over 200 miles on one tank of hydrogen.

Because hydrogen is not very energy dense, a tank of a given size will only provide around 70 percent of the range of a gasoline tank. A way to address this issue is to compress the hydrogen in a metal tank in the same way that other gases are compressed and stored. As you can imagine, these tanks are heavier and require a lot more safety features because compressed gases are dangerous. In a collision, the tank may potentially turn into a deadly missile. One of the key challenges facing HFV designers is how to store pure hydrogen on board. Following are some possibilities:

- **The "Cryo" fuel tank,** developed by BMW. Hydrogen is stored in a special well insulated cryogenic (extremely cold) tank maintained at minus 400 degrees Fahrenheit. The tank is stored behind a fortified bulkhead located directly behind the rear seats (it's big and the trunk is small, which is a practical limitation). There are potential dangers associated with this scheme: In an accident, for example, the tank may rupture and the tremendously cold hydrogen could literally freeze an occupant. Or the tank could also explode when the hydrogen is mixed with oxygen (but great pains have been taken to ensure this doesn't happen). Still, it's a daunting thought and the likelihood of this kind of system gaining commercial acceptance is remote.

- Metal *hydrides* (compounds consisting of a combination of metal and hydrogen) can store hydrogen, which can be liberated at a controlled rate for use as a fuel.

- **Gas-on-solid absorption technologies** store hydrogen in extremely small carbon manifolds called nanotubes. Hydrogen gas condenses within these nanotubes at densities comparable to those of metal storage tanks, but with none of the high-pressure dangers of storage tanks.

    More research and development needs to be done on both the metal hydrides and the gas-on-solid absorption technologies before widespread acceptance will be seen.

Hydrogen tanks are potentially dangerous and require a lot of servicing, maintenance, and inspection. There are very few qualified technicians who can do this right now.

Fuel cell vehicles

A fuel cell vehicle (FCV) is basically an electric vehicle that incorporates a fuel cell in place of, or in addition to, a battery bank.

The electric motor powers the drivetrain in the same way that a conventional all-electric vehicle works (refer to Chapter 18). The electric motor gets its power from either the fuel cell, the battery bank, or both at the same time depending on load conditions. The fuel cell can recharge the battery bank when the vehicle is stationary (even when the vehicle is turned off and is parked). If battery banks are used with the fuel cell, when the hydrogen tank runs dry, there will still be enough power available to run the vehicle (just like a standard hybrid vehicle).

These vehicles can also be plug-in type, so that the batteries can be recharged via grid power. Recharging via the grid might be necessary, for example, when hydrogen is in short supply or unavailable.

The overall pollution levels achieved with FCVs is only a fraction of that achieved with internal-combustion engines. Typical efficiencies are better than fossil-fuel-powered engines as well. However, efficiencies are not as good as with hydrogen-fueled vehicles (refer to the preceding section).

In addition, the maximum operating range of a hydrogen FCV is less than that of an equivalent internal-combustion engine vehicle. However, if methane or propane is used to derive the hydrogen, comparable range performances are achievable. As with all-electric vehicles, servicing may be difficult because the infrastructure is limited.

A Hydrogen Future

Of all the alternative strategies considered today (and explained in this book), hydrogen fuel cells offer the most promise. While other options can offer some benefits (reduce greenhouse gases and pollution, offer energy savings and efficiency gains, or create the power output required by the modern world, for example), hydrogen, once fully developed, can offer *all* these advantages.

The problem with hydrogen fuel cells is that the economics are not currently competitive. Operating fuel cell vehicles costs more than operating conventionally-powered vehicles and requires tenacity and determination because the resources are so much more difficult to obtain. Today, if you buy and operate a fuel cell-powered vehicle, you have to pay more for the vehicle up front, and you have to plan ahead for fill-ups because you simply can't obtain the required fuel as easily as you can obtain a gallon of gasoline. It takes a hearty commitment to drive a fuel cell-powered vehicle.

Until recently, global warming was considered a problem of the future. Unfortunately, humankind has shown throughout history that solving future problems is rarely a priority (never do today what can be put off until tomorrow), and as a result, developments in energy technologies have languished. Yet as the effects of global climate change have become more apparent, many more people recognize the real threat it poses.

For this reason, people are willing to spend more to go green and to make extraordinary commitments to do so. All sorts of alternative energy schemes, like solar and wind power, that aren't cost competitive with fossil fuel sources are proliferating. As more people demand the technology (and are willing to pay the higher price for it), the technology will become more common and prices will come down. As prices get lower, more people will get on board, and eventually, broad acceptance of hydrogen technology will materialize.

Chapter 21

Exotic Propulsion Systems

▶ Promising propulsion systems

▶ Examining fully functioning systems in use today

As far as sustainable, nonpolluting, readily available, and cheap go, you can't beat the propulsion system used by Fred Flintstone and other drivers in Bedrock. Alas, most people aren't willing to hoof it from place to place while hauling the car along, too. Fortunately, there are other alternatives.

In this chapter I describe some of the more exotic transportation scenarios being developed to address the pressing need for more efficient and less-polluting means of moving people and things around. Some of these schemes are still on the drawing board, but some are already being used extensively (nuclear-powered ships, for instance).

Supercapacitors

Energy can be stored in capacitors, which are very simple electrical devices used in literally every single electronic device on the planet. Here's how a capacitor works:

1. **A battery connected to the capacitor deposits a charge onto the two opposing plates.**

2. **When the battery's disconnected, the charge remains on the plates.**

3. **When a load (a motor, light, or electronic device) is connected to the capacitor, the charge flows through the load, providing power.**

In short, a capacitor is a device for storing energy; it doesn't create or produce power; it simply stores it (like a battery).

Supercapacitors are simply immense versions of the same, simple concept. There is no theoretical limit to the amount of energy they can store. Energy densities (the amount of energy stored per unit volume and weight) can be increased by manufacturing the capacitor in such a way as to prevent leakage (over time, every capacitor self-discharges due to imperfections).

The amount of charge that a capacitor can store is directly related to the surface areas of the opposing charge storage surfaces. Electromechanical double-layer capacitors (EDLCs) have been developed using carbon electrodes with extremely high surface areas of up to 2,000 square meters per gram of weight. These are used in conjunction with an electrolyte (the material located between the two electrodes) of aqueous acid such as sulfuric acid. Organic-based electrolytes, which are safer for the environment, have also been used successfully.

In practical terms, supercapacitors can replace batteries. But in all likelihood, supercapacitors won't necessarily replace batteries. Instead, they'll be used in conjunction with batteries. Because capacitors can deliver a great deal of power in a short period of time, they are ideal as power boosters, for when an electric car needs to accelerate rapidly, or needs to crest a steep hill. Supercapacitors are also ideal in hybrid vehicles, displacing a good deal of the current battery capacities with safer, lighter energy storage means.

A key benefit of supercapacitors is that charging them takes much less time (on the order of 1 percent of the time) than charging a battery. In fact, you can charge a supercapacitor in about the same amount of time you can fill a gas tank, making these vehicles far more practical, in real world applications.

One concern of capacitors is safety. Capacitors are capable of discharging tremendous amounts of power in a very short period of time. In an accident, a capacitor powered vehicle could severely shock the occupants, or emergency and support personnel. These problems can be overcome, however, by suitable electronic control circuits.

The widespread application of supercapacitors for use in powering vehicles is years away, but the technologies are improving quickly, and venture capitalists are entering the game, which means that they believe in the market potential.

Solar-Powered Vehicles

Solar PV cells generate electricity. When PV modules are located on the roof of the vehicle, or for that matter, literally anywhere on the vehicle's body, in theory, the electricity generated can power the vehicle. The more sunshine, the better. Solar cells have already been used to power vehicles — even aircraft. However, these vehicles are very lightweight and don't offer much by way of high-end speeds. As the sole power source for vehicle propulsion, solar has a lot of disadvantages:

✔ **Solar cells simply can't provide enough energy per time to offer performances that people take for granted.** Current electric cars consume around 260 watt-hours per mile, while solar cars are only capable of delivering around 25 watt-hours per mile. That's energy frugality at its best, and it precludes all creature comforts like nice seats, heating, radios, and so on. If you have infinite patience, though, then solar propulsion may be just the thing.

✔ **These vehicles only operate when the sun is shining.** The more sun, the faster the vehicle will go — which isn't exactly how people want to drive their cars, although it often seems like the only explanation for how some people drive.

As solar cells become more efficient (that is, can produce more electrical power per unit of area), however, solar cars will become more practical. When the first solar cells came onto the market back in the 1960s, their efficiencies were only on the order of 6 percent (6 percent of the total radiation energy falling onto the collectors was converted into useable electrical energy). Solar cells are now available at over 30 percent efficiencies, and this number will climb to over 50 percent in the next decade, particularly with the strong interest in solar power that the world is experiencing.

Until then, solar cells can also be used in other ways to improve the energy efficiency or environmental friendliness of vehicles:

✔ **To charge batteries in an electric car or a hybrid:** Many cars sit in the sunshine all day long while the owner is working, and by building enough solar collectors into the car's body, particularly the roof, there may be enough solar power collected during the course of a sunny day to allow short-distance commuting completely derived from the sun. Or the solar panels may be used to charge supercapacitors (see the section, "Supercapacitors," earlier) instead of batteries, making the entire process very environmentally friendly.

✔ **To augment the support systems like air-conditioning (which is needed only when the sun is shining brightly) in all-electric vehicles:** Air-conditioning loads are heavy; they sometimes take up more than 40 percent of the total power a vehicle is consuming, particularly a small, electrically powered vehicle.

✔ **To make internal-combustion engines more efficient.** Over 50 years ago, MIT scientists demonstrated what's called *thermophotovoltaics* (TPV). When illuminated by an infrared light (which can be obtained from a very hot source), solar photovoltaic cells can produce up to 60 times more power output than when simply illuminated by sunlight alone. Combusting fossil fuel provides a very hot flow of infrared radiation. Because there are a number of heat sources in a conventional auto, placing photovoltaic cells in strategic locations could recoup some of the wasted heat, making the auto much more efficient.

Solar power for vehicles will entail a slow, evolutionary process over the next few decades as solar becomes more efficient, and the cost comes down. Look for solar panels on cars in the next few years, but only in minor roles. Twenty years from now, solar may play a major role.

Flywheels

Most people know what a flywheel is: a heavy weight spinning very fast, like a gyroscope. Conventional autos incorporate flywheels in their engines to even out load conditions and provide stability to the engine (which is undergoing periodic jolts of momentum from the individual pulses from each cylinder firing). Flywheels are also used to create the smooth, even motion of a ceramic potter's wheel.

What most people don't know is that flywheels are really just energy storage devices. And it's possible to design flywheels to store a tremendous amount of energy, enough to power a vehicle.

In an electric vehicle, a flywheel would be used to augment the low power outputs of the typical battery bank/electric motor combination. Here's how it works: The flywheel spins during normal driving conditions. Then when a big burst of power is needed, the energy in the flywheel is called upon to provide that burst of power. This operation is similar to that of the supercapacitor (for more on supercapacitors, see the section, "Supercapacitors," earlier), but it's mechanical instead of electrical. (Flywheels are often referred to as *mechanical batteries.*)

Because of the huge amounts of power that are available with a flywheel, accidents are a problem. The newest flywheel technologies solve this problem by using composites that shatter into infinitesimal pieces upon impact. The pieces are so small as to be considered a liquid, and the energy of the flywheel is disbursed without harm to the vehicle occupants.

Regenerative braking systems (those that recapture a vehicle's kinetic energy when the vehicle slows down) work very well with flywheels. When a car stops, or goes down a hill, the recouped kinetic energy is simply translated directly into the flywheel. This is far more efficient than the electrical regenerative braking systems used on current hybrids because those systems are comprised of mechanical-to-mechanical energy.

Flywheel technologies are already being used in some prototypes, and the promise is good enough that research-and-development money is being dedicated on an increasing scale. In the next decade, flywheels will be incorporated into hybrid vehicles on a routine basis.

Hydraulic Accumulators

The hydraulic accumulator is a means of storing mechanical energy, much like a tank of compressed gas. A hydraulic accumulator can store far more energy than a simple tank of compressed gas, however, and it's much safer. It also takes up much less volume, and so the technology is being developed to power vehicles. It operates in a similar fashion to the flywheel (see the section, "Flywheels," earlier). It stores energy during braking and going down hills, and can then feed this energy back into the drive system of the vehicle, as required.

The accumulator stores energy by pumping oil at high pressure into a pressure containment vessel. The oil now holds a lot of potential energy that can be drawn back out at will, at a high rate of speed.

The accumulator is not capable of powering a vehicle for any distance, but it's perfect for augmenting small-sized electrical power systems. Accumulators will never be the sole power source for a vehicle because they only store energy; they don't create it. But they may replace batteries in hybrids once the technology has matured.

Magnetic Levitation Trains

Mass transit, which includes buses, trains, vans, and so on, is much more efficient at moving people and loads than are individual autos. One of the best ways to solve the energy problem is by building convenient, accessible, on-time mass transit systems, in particular trains. The most efficient of all mass transit vehicles is the magnetic levitation train. In some countries, there are already a good number of magnetic levitation trains, and plans are on the board to build many more. Europe and Japan already have a number of magnetic levitation trains, and there are a few operating in the Eastern corridor of the United States. California is finalizing plans to build a high-speed magnetic levitation train between the Bay Area and Los Angeles.

Using magnets

Magnetic levitation uses magnets to suspend objects in midair. You've probably seen children's toys that suspend a metal ball over a platform loaded with magnets, appropriately placed. The same phenomenon is used to suspend a large object like a train. The result is that friction is virtually eliminated.

Of course, magnets have north and south poles. When you put a south pole next to a north pole of another magnet, they attract and snap together. When you try to push a north pole at the north pole of another magnet, the two magnets repel each other. When the magnets are very strong (electrical magnets can be made as powerful as desired), the attraction/repulsion can be very forceful and a very large weight — say an entire train complete with passengers, baggage, and so on — can be suspended in midair, with little expenditure of energy.

The lower system (the "track") is fixed into place, and the upper system (the train) levitates over the lower system. Feedback systems keep the magnetic elements properly aligned with respect to each other. (When magnets repel each other, they wobble off center.)

Superconductors — metallic materials at very low temperatures that can literally channel the flow of current without resistance — make magnetic levitation practical and economical because magnetic fluxes (magnetic force fields) are entirely absent from the superconducting metal (a phenomenon referred to as the Meisner effect, after the discoverer). Because of the Meisner effect, the magnetic forces produced by a superconducting magnet can be extremely large, with very little expenditure of power. An electrical current imposed onto a superconducting magnet can flow around and around for a long period of time without dying out. A superconducting magnetic medium creates a much stronger repulsive force for a set magnetic level than ordinary magnets do.

Bullet trains

Trains that use superconducting magnets to levitate the train cars are called *maglev*. There is no contact between the cars and the track. The gap between the train cars and the track is around one inch, which isn't much but that's all it takes. The only mechanical friction that occurs is due to air resistance, and that can be minimized by suitable aerodynamic designs. Maglev trains are often called "bullet trains" because of their bullet like shape and the fact that they can go extremely fast (over 200 miles per hour) with great efficiency.

There are two different types of levitation schemes in use, the monorail and the wraparound track. In the *monorail design,* the cars are attached to bearings that essentially wrap around the track. In the wraparound design, the track wraps around the bearings attached to the railroad cars. The most common arrangement is the monorail design.

In both systems, vertical magnetic fields levitate the cars, while horizontal fields keep the cars aligned properly on the track to avoid friction. Acceleration and braking are accomplished via a set of linear motors that use magnetic fields to interact with the stationary, track magnetic fields. There is no mechanical contact between any part of the train cars or the track.

Pros of maglevs

Maglevs have a number of advantages as an efficient, comfortable transportation alternative:

- They're very quiet, even at high speeds. Because the train and track never meet, there are no grinding or cranking sounds at all and the ride is quiet and smooth. At top speeds, the countryside literally flies by and commute times are reduced considerably.

- Maglevs are entirely electrical, which means that when they are run using nonfossil fuel alternative-energy sources to produce the electrical power they are very clean for the environment. And because they don't need fuel tanks, they're lighter and safer than conventional transportation schemes.

- Maglevs can compete with airlines not only because of their speed, but also because of the logistics of boarding and unboarding: You can park a few yards away from the maglev boarding station and hop aboard with relatively little hassle. Even in long-distance travel, the total time taken to ride a maglev can be less than that for an aircraft which travels much faster. For those who have flying phobia, maglevs are the solution.

Cons of maglevs

Maglevs are very expensive to build compared to other trains. They are economical only when filled to capacity and used all the time. Maintenance is expensive, although infrequent. Other areas of concern include the following:

- Passengers are subjected to strong magnetic fields. There's been some concern about whether this exposure poses health risks. Research is being done to determine the answer, but so far no hazardous effects have been recorded.

✔ Power outages in the grid mean the train will come to a halt. This could be dangerous when the train is moving very fast, because the levitation will suddenly disappear. Safety systems are installed to alleviate this problem, but there is still a higher safety risk than that for conventional trains because accidents are potentially more catastrophic.

✔ High winds can push the train around, causing momentary contacts between the levitation elements, which may be dangerous.

Nuclear-Powered Ships

The main application for nuclear-powered ships has been in the military arena, principally submarines and aircraft carriers. However, with improving nuclear technologies and an increased concern for fossil fuel usage, nuclear power is now being considered for a wide range of commercial vessels as well.

Nuclear fission is the type of power plant used in nuclear-powered ships (you can find a detailed explanation of nuclear fission in Chapter 8). In a ship, the fission reaction is carefully controlled within the nuclear reactor containment vessel (highly shielded to prevent radiation leakage). Heat from the reaction is channeled into the boiler, which powers a turbine/generator combination that produces electricity. The electricity then powers a huge electric motor that drives the propeller. Another turbine/generator provides additional power for the ship's support systems. Nuclear powered ships are all-electric.

Efficiency and environmental impact

Nuclear power is far more efficient in terms of providing miles per unit of raw fuel than conventional, combustion-driven power sources. Plus, these ships emit no greenhouse gases. What nuclear-powered ships do emit is radiation, which can be considerable if the containment vessels are inadequate. This affects not only the crew, but sea life beneath the ship. Radiation trails in the water can last for a long time.

Nuclear, of course, has its detractors in the green movement. The political implications of building a commercial, nuclear-powered ship are onerous. However, with the increased concern over global warming, nuclear-powered commercial vessels will surely be built in the next few decades.

Performance and maintenance

Nuclear-powered ships can go a lot faster than conventional ships, with much less weight in terms of fuel. In addition, nuclear reactors don't need an oxygen source, unlike combustion engines. This makes them perfect for submarine applications. (Conventional submarines need to surface periodically to draw in oxygen while nuclear-powered submarines may remain submerged for months at a time.)

Maintenance, although itself a big job that requires a lot of high-tech equipment and often needs to be done in port, is needed less often.

Safety issues

Nuclear power is a proven, reliable, and safe technology. A lot of problems can occur on conventionally powered ships, like fuel spills, fire hazards, and the like. While problems can occur on nuclear-powered ships (read on for more), on balance, they're safer than conventionally powered ships, a fact that the record bears out. Why? Because of the highly advanced safety technologies and procedures that nuclear protocol requires.

While the risk is small, thermal meltdown can occur. There have been several instances of submarines (Russian) experiencing catastrophic nuclear reactor accidents. But like the Chernobyl nuclear accident, the Russian problems have been more a function of institutionalized laxity than fundamental technical inferiority of nuclear power.

As with all nuclear materials, terrorists would love to get their hands on some samples that they can use to blackmail or terrorize a community.

For the truly adventurous: Bicycle transport

There is no more efficient and economical transport system than a good, lightweight bicycle, as has been proven by numerous studies conducted over the last 300 years. Bicycles have two wheels and no onboard power source other than the rider, who is fueled by a good breakfast of healthy nuts, grains, and fruit. And because food is available virtually everywhere on the planet's surface, there's no problem refilling the energy source.

(continued)

(continued)

Advantages of bicycle transport:

✔ Lowers blood pressure and cholesterol levels, increases overall health and well being of the rider, and increases sex drive and longevity.

✔ With tandem bikes, increases quality of interpersonal relations, unless the rider in front refuses to listen to the rider in back or the rider in back refuses to pedal.

✔ Is fun, when done under the right circumstances (see the preceding bullet).

✔ Decreases our dependence on foreign oil and reduces global warming gases.

Disadvantages of bicycle riding:

✔ May take a long time to get wherever you intend to go and, if your energy runs out, may not get there at all (unless you camp beside the road for a few hours while you recharge your power source).

✔ Can result in ridiculous displays of spandex and decals.

✔ Causes large thigh muscles which may make it necessary to purchase specially designed clothing, particularly tight pants.

✔ Can result in several moments of confusion, disbelief, anger, and finally self-pity as you walk home because you didn't have a good lock.

✔ Is an irritant to conventional transport drivers when you don't yield the right of way, which on a bike, you're often tempted to do.

Part V
The Part of Tens

The 5th Wave   By Rich Tennant

"They're solar panels. We're hoping to generate enough power to run our tanning beds."

In this part . . .

This would not be a *For Dummies* book without these handy chapters of ten. Because alternative energy is widely misunderstood, I dispel the most common myths and provide a list of the best ways to invest in alternative energy so that you can begin to make a difference, right now, today.

Chapter 22

Ten or So Myths about Energy

A lot of stubborn myths abound when it comes to energy. By reading this book, you can gain an appreciation for the subtleties inherent to the subject of energy and the knowledge to separate the wheat from the chaff. The fact is, while energy itself is an easy concept to understand (it's simply useable power), how energy works — the physics of it — can be pretty slippery. Here's an example of how the principles governing energy's characteristics isn't intuitive: After you use energy, it's gone, right? Wrong. Add to this the media's tendency to exaggerate problems because it sells copy and downplay the good news because it doesn't, and you have fertile territory for mythology. In this chapter I explore and attempt to debunk some of the more prevalent myths.

The World Is Doomed to a Gloomy Future

The current global warming forecasters warn of a dismal world, with rampant starvation and global population displacement — unless someone steps in now to do something about it. The little glimmer of hope that quick action can make a difference is often doused by two things: the common perception that humans are, by and large, incapable of caring about things that don't present an immediate danger and by how ill-prepared the world seems to be to address the problems. When you read a newspaper or watch the news, it's easy to get hung up on the negatives, and it seems as though the only environmental news is bad news.

But people do care and, when motivated, can do remarkable things. Don't let anybody tell you any different. Problems get solved, perhaps a little later than they should, and perhaps with a little more fanfare than they should, but they do get solved eventually.

Case in point: Back in the '60s, there were a lot of environmental problems that have since been cleaned up due to the fact that people care and can marshal the resources necessary to act for the good of the environment and their communities. I remember, in particular, a photo shoot of Lake Michigan before and after cleanup legislation was enacted. The before photo showed a human hand that had been dipped into the lake — it was covered in gory slime. Ten years later, an identically staged photo showed a hand with little more than clean water.

As far as the global warming, people are starting to take great strides in the right direction. They're spending money on new cars that get better gas mileage. They're making changes in the way they consume energy and demanding that the businesses they rely on also go green. They're calling their representatives and sharing their concerns. Even if you've done nothing else, you've bought this book because you want to understand the options that are available or on the horizon. (Personally, I think people should buy more copies of this book and hand them out to all their friends.)

Bottom line: The current dilemma will be solved because people are basically good and want to do the right thing.

We Can Legislate Our Way Out of the Mess

Many people believe that it's the government's responsibility to legislate environmental and energy laws that will solve the country's addiction to fossil fuels and encourage conservation and efficiency. While government must lead the way, it's the responsibility of every private citizen to engage solutions on whatever level they can contribute — even if this means entering into the political process itself so as to get the system moving the way it should. Relying on government is a form of passing the buck, and it doesn't really work very well anyway.

Energy efficiency at home is the best place to start. Using less energy is very easy, for most people. It begins with education (see my book *Energy Efficient Homes For Dummies* [Wiley] for a good start). Understanding how and why you use energy, and what it accomplishes for you is imperative. In my experience, most homes and businesses can cut their power bills by over 20 percent without experiencing any appreciable loss of quality of life.

Every little bit helps, as the saying goes. And every thousand-mile journey begins with the first step. Just remember that it's up to you, not your government.

The Electric Car Was Deliberately Killed

Conspiracy theorists claim the big auto companies could be providing consumers with automobiles that get over 100 miles per gallon but they don't because it will affect their bottom line. In the same vein, there is a lot of talk about how the big auto companies killed the electric car simply because they're cheaper (being smaller), and so the profit margins will be less.

The fact is, all successful companies basically provide their customers with what the customers want. Americans drive massive SUVs and trucks because that's what they want to drive. If Americans wanted electric cars, they would have them. If they wanted mopeds, they would have them. (Some things they get whether they want it or not, like ineffectual government.)

The electric cars that were introduced back in the early '90s were inferior for a number of important reasons, the most important being their limited range and how long it took to recharge the batteries (you can read more about these early cars in Chapter 18). When electric cars make sense and can meet the performance standards of modern drivers, people will buy them and the automakers will build them exactly the way people want them.

As an interesting side note, consider what has happened to the American auto makers. They did not anticipate that consumers would want small, fuel-efficient cars. The Japanese car companies have done a good job of developing and providing these, so American manufacturer's market share has dwindled. It's very simple; you give the consumer what they want and you make money.

Now as to whether Big Oil kidnapped the inventor of a 100 percent efficient engine that operates only on water . . .

Conservation and Efficiency Can Save the World

No matter what you may think, conservation and improved efficiency don't lead to less energy consumption; they actually lead to more. The fact is, the more efficient a technology is made, the more people use it. When cars get better gas mileage, people drive them more and make up for the difference. When televisions and appliances are more efficient, people use them more. When fossil fuel extraction technologies become more efficient, the cost of energy goes down and people use more energy. Efficiency has rarely resulted in a decrease in consumption. (Go to Chapter 7 for a complete discussion of this phenomenon.)

Conservation and efficiency work on a micro, or household, level. When people are conscientious about how they use energy, they can easily use less and the effects can be profound. But on a macro scale, conservation and efficiency only work when there is an economic incentive. For that reason, government needs to apply higher taxes to energy in order to make conservation and efficiency work on a broad scale. This is why European cultures use less energy, per capita, than North America. European energy taxes are much higher than American taxes. Yet people in Europe do not have a diminished standard of living compared to ours.

There's an Energy Shortage

Energy can neither be created nor destroyed. It can only change forms. I explain this in some detail in Chapter 3. When fossil fuels are extracted from the ground and burned, energy merely changes form from the liquid crude potential energy form into heat, which then changes form once again into motive energy (energy which drives machinery). To say that energy is used is entirely incorrect.

That's why there is no shortage of energy and there never will be. There may not even be a shortage of fossil fuel energy resources for another 100 years or so. The problem is not that the country or world is running out of energy, it's that the current use of fossil fuel energy is causing too many problems. Pollution is at the forefront, but Americans also experience political and cultural problems when too much fossil fuel is used. There is certainly no shortage of solar energy, nor wind energy. Over 35,000 times more energy than humans use falls onto the earth every day from the sun. If we could just convert a small portion of that into useable electrical energy, we'd be set.

There is virtually an infinite supply of energy available to humankind. The question is how the various forms of energy are transformed into work that we wish to accomplish. And *that's* where big improvements can be made.

Nuclear Power Plants Are Ticking Time Bombs

Nuclear power plants cannot explode like an atomic bomb. The physical process is entirely different between a nuclear reactor and a nuclear bomb. Reactors are based on fission while bombs are based on fusion. Bombs rely on chain reactions and the fuel is spent almost instantaneously. Nuclear reactor rods are consumed slowly, and there are a host of safeguards to make sure that meltdowns and other accidents cannot occur.

The containment structures built around the nuclear reactors in the free world are designed to withstand tremendous explosions and still keep the reactors intact. Three Mile Island proved this. (The Russian reactor at Chernobyl that caused so much environmental and health damage was intentionally built substandard. Its failure was not a reflection of nuclear reactor technology.) France has proven that a well-designed, well-regulated nuclear reactor system can be a viable and effective source of energy. It's too bad the United States didn't copy France's lead; otherwise, America's political and economic situation would be far different today.

For more about nuclear power, head to Chapter 8.

Electric Cars Will Displace All Others

While electric cars offer many advantages over fossil fuel–powered cars, they are not destined to become the sole type of transportation of the future; nor do they solve all the problems inherent in fossil fuel use. The first and foremost problem with electric cars is that they need to be plugged in and recharged on a periodic basis. While they don't emit noxious gases themselves, the power plants that provide the electric power do. And in the United States, over half of its electrical power comes from coal-fired power plants. Coal is one of the dirtiest sources of energy, and there are no truly promising solutions that will eliminate this problem. There is talk of clean-burning coal plants, but what this basically means is that the power plants will be cleaner than they used to be, not that they'll be clean. Coal may never be truly "clean," compared to alternatives.

The most sensible program would be to use a large number of nuclear reactors to provide the electricity to recharge electric cars. But nuclear reactors take decades to build, and this rosy scenario of nuclear-powered electric vehicles will take at least another 30 years to come to fruition.

Anything Labeled "Green" Is Green

It's the in thing these days to claim that your company is "green." The race is on to see who's the greenest. Business used to be all about being lean and mean; now they're about being green.

You'll hear companies proclaim that they are "carbon neutral," meaning that on net, they emit no greenhouse gases, or see advertisements on television and in the media lauding their greenness on a daily basis. Even stock brokers are racing to see who is the greenest (touting things like, "We don't use any paper in our offices!").

But when you read the fine print, you'll often find that there are some significant exceptions. For instance, a manufacturer may claim that its carbon footprint is zero, but this may be true for only a small portion of the manufacturing cycle. No mention is made of transporting the wares to the consumer, which takes a lot of gasoline. Or no mention is made of the fact that many of the parts are made in a country, like China, where environmentalism is nearly nonexistent.

So when you hear claims of "greenness," keep this in mind: There is no governing regulations with regard to the word "green." Anybody can use the word, and if their claim is a complete fabrication, there are no penalties. If you are looking for green companies to do business with, look deeper than the surface and find out just what is actually green about the company's products and processes. If you think they're stretching it, write a letter to a local newspaper and expose them. Then they'll have to make more effort and clean up their act. Or else use more green paint.

Waste is Bad

Whenever energy changes form, waste is an unavoidable result. When gasoline is burned, there is a good deal of waste, and there's nothing that can change that. When a solar panel converts sunlight into electricity, a great deal of waste results. Every energy process results in waste. The more refined the energy process, the more waste. Creating the refined energy needed to drive a modern computer chip entails a huge amount of waste, and that will never change. The more sophisticated our computers become, the more waste the usage of those computers will entail. Period. (You can read more about this concept, called *entropy,* in Chapter 3.)

We need energy for many different reasons. We use energy to provide our food supply, the most basic of all needs. We use energy for transport, for heating and to provide us with the myriad forms of entertainment that we take for granted. Whenever we use energy, we waste a lot more than we use. That will never change simply because it can't. To say that waste is bad is to say that energy usage is bad.

The historical record states unequivocally that societies progress along with the sophistication of their use of energy. The difference between the third-world and first-world countries is very simply ascribed to the fact that first-world countries use much more energy and as a result have a far greater economic output. They also waste a lot more energy than third-world countries as well.

If our economy is to grow and prosper, we will need to use more and more energy. And we will, as a consequence, waste more and more energy.

Waste is good! Rejoice in waste in all its forms.

Solar Power Can Provide All Our Electrical Needs

Solar power is an excellent investment for those lucky enough to live in conducive climates, such as the Southwestern United States. As technological improvements take hold, and the cost of solar decreases, more and more people will be able take advantage, but solar power will never be able to provide all of the country's energy needs.

In order to power New York City with solar power, for example, an area of solar photovoltaic panels twice the size of New York City would be required. Where are these panels going to go? Over the city, or in somebody else's backyard? Way off yonder where nobody can see them?

In addition, solar power works only when the sun is shining (it don't work where the sun don't shine, as they say in the industry). Unfortunately, most residential and business energy is consumed when the sun isn't shining, for instance, on a cold winter day.

Finally, solar power is not economical unless government subsidies help reduce the cost. This is not likely to change in the next decade. Solar power is very good for certain applications, but it's not going to take over the world for a long, long time.

Burning Wood is Bad For the Environment

Like all biomass, wood is carbon neutral, which means that the carbon contained in the raw source came from the atmosphere (while the plant material was growing), and when the biomass source is burned that carbon simply reenters the atmosphere from whence it came. In other words, a chunk of wood left to rot on the forest floor releases as much carbon dioxide into the atmosphere as that same chunk burned in a wood stove. It's a closed loop, and so wood burning does not contribute to global warming.

The problem with wood burning is that most of it is done improperly or very inefficiently. Most wood stoves are old and leaky, for example, and the wood burns too fast. And most wood stoves do not deliver heat into the interior spaces of the home or business nearly as efficiently as they could. Most of the heat goes right up the chimney and out into the great outdoors where it's not needed at all.

The solution is to use high quality, modern equipment to burn the wood. This makes a lot of economic sense because the cost of heating is drastically reduced when efficiency is increased.

If Global Warming Isn't Our Fault, We're Off the Hook

Since I'm an open-minded guy, I must conclude this chapter with an open-minded analysis of some hard facts. Thirty years ago, global cooling was all the rage. Magazines loudly trumpeted the end of the world via a creeping ice age predicated by mankind's fossil fuel–burning emissions. Leading scientists warned that if we failed to curtail our use of fossil fuels the world would soon freeze over. There was scientific consensus at the highest levels.

How things have changed. Now we're inundated with warnings about global warming, and the evidence is crystal clear. But the evidence for global cooling was also crystal clear. The fact is, Earth has undergone over 600 warming and cooling cycles. Humans had nothing to do with these cycles. So it may be that the current warming trend is simply Mother Nature acting fickle, as she always has.

Regardless of where you fall on the "is it or isn't it real" question, a big question still looms. What should we do if global warming is real but is not being caused by our greenhouse gas emissions? Should we forget all this alternative-energy nonsense and keep blasting away with our fuel-hungry leaf blowers? Should we continue to drive our Hummers back and forth to the mailbox across the street because we're too lazy to walk?

In my view, the Golden Rule applies. We should simply treat Mother Nature the same way we would like for her to treat us. We've been doing a poor job of that lately, and I think we can improve. And that means pursuing alternative-energy solutions that protect the environment *and* make economic sense.

Chapter 23

Ten-Plus Ways to Invest in an Alternative-Energy Future

In This Chapter

▶ Changing habits

▶ Looking at where you live

▶ Figuring out where and how much

*B*y now you're chomping at the bit to do something tangible. You've found out a lot about efficiency and alternative energy and have some ideas how you can make an appreciable impact. In this chapter, I present some of the best and most direct options for you to consider. The list is certainly not exhaustive, and everybody can contribute on his or her own terms. But I've tried to winnow the options down to those that make the most sense and impact.

Performing an Energy Audit of Your Home or Business

An energy audit is a formal process for evaluating your home or business to see precisely where, how, and how much energy you are using. You can hire a professional auditor (look in the yellow pages or on the Internet), or you can perform the audit yourself. It's not hard; it doesn't take a lot of time; and it's well worth the effort.

In the process of performing the audit you'll learn a lot of things that never occurred to you. For instance, how much propane do you use in your family's barbeque? How much gasoline do you use to mow your lawn or blow the leaves off of your driveway? How many batteries do you use? How much water do you use? Do you recycle, and how much energy are you saving in the process?

Even more than the actual hard numbers that you end up with, you get a broader picture of just how much and which types of energy you use — info that may both surprise you and present improvements you can make immediately to reduce your energy consumption. As a professional auditor of both businesses and homes, I've found that most people can save between 10 and 20 percent of their power bills without sacrificing much in the way of quality of life. Many people can save over a third of their power bills by simply changing habits and rearranging the way they use energy. You don't need to buy new appliances (although you may want to after you see how much that old refrigerator is costing you).

To perform an energy audit, you want to find out how much energy you're using and how much it costs (you can find complete instructions in my book *Energy Efficient Homes For Dummies* [Wiley]):

✔ **Get a history of your power bills.** You can request records from your utility company, or you can retrieve old bills from your file cabinet. Simply by studying the trends you'll learn some important things like how much your air-conditioning costs compared to your heating.

✔ **Make a list of the appliances in your home or business that use energy and evaluate how much energy each appliance uses.** Most appliances have labels that tell you how much power they use. Measure the amount of time each appliance is on and multiply that by the power consumption to get a loose idea of how much each appliance is using. Don't forget to include the cost of space heaters, swamp coolers, ceiling fans, and the like.

✔ **Evaluate how much that energy costs.** On your power bills the rate schedule will be defined. Multiply the amount of energy by the cost, and you can determine how much each appliance is costing.

The bottom line is that the more information you can compile, the better your efforts will be focused to reduce your power consumption.

Installing a New Stove

If you burn wood in your home, chances are you are using an old, cast-iron stove that may have sentimental value but is inefficient and highly polluting to boot. The vast majority of wood stoves are over ten years old, and that means they're substandard compared to today's technology. New wood stoves use advanced combustion technologies that make the burning process up to five times more efficient than old stoves. This means that it'll cost a lot less to heat your home and reduce the amount of pollution you're emitting by about ten times. You'll be feeding wood into the stove much less as well, not to mention the chopping and carrying and stacking and spiders and backaches.

Even if you don't switch to a new stove, you can probably use your existing stove more efficiently. I describe how in Chapter 14. You'll save money and help the environment at the same time. It may take a little effort to learn better techniques, but once they're in place you'll benefit forevermore.

Even better, change from wood to natural gas, which burns much cleaner because the combustion process is so highly regulated. When you burn wood, it's nearly impossible to get the same emission efficiency as a good gas stove because wood burns differently depending on how much of a given chunk has already burned (when you first light a wood stove, the burning is very inefficient, and you can't get around the starting phase unless you simply burn in perpetuity). A good gas fireplace emits far less pollutants per unit of heat produced than a wood stove.

However, I will have to concede that burning gas can be more expensive than burning wood (especially if you have a source of free wood). I switched from wood to gas and my overall cost of heating went up around ten percent. But it's worth it. No more spiders or dirt. No more chopping wood, or stacking or carrying or cleaning out the filthy ashes. With a nice new gas fireplace all you do is press a button and you have heat (sounds kind of lazy, doesn't it?). The new gas stoves look very nice, and the flames don't seem artificial at all. (Forget electric fireplaces — they look weird and consume a lot of electricity.)

Best of all, install a stove that burns biomass-produced natural gas. Or install a corn or other biomass stove that's extremely efficient. Check out Chapter 13 for more details on biomass stoves.

Using Solar

If you're in the right climate, solar PV can be a great investment, and it helps the environment more than any other alternative-energy source you can employ. Solar power entails no greenhouse gases at all (aside from the invested energy consumed in the manufacturing process of the panels and equipment). And there will never be a shortage of solar power. See Chapter 9 for more details on the great advantages of going solar.

Governments are subsidizing solar PV systems to the tune of nearly 50 percent of system cost, in some parts of the country. You can obtain financing from some utilities, which makes the out-of-pocket expense minimal. And once your solar system is up and running, your power bills will be very small. Solar customers love their systems; just ask somebody who has solar. They'll tell you the investment was well worth it. In fact, a vast majority of solar customers will tell you that they wish they had installed a larger system. That's not a cheesy sales hook, it's a fact.

Another good solar investment is installing a solar hot-water heater on your roof. The typical home consumes anywhere between 15 and 20 percent of its total energy budget on heating water for use in dishwashers, washing machines, sinks, and showers, among other things. A solar water heater can displace up to 80 percent of this cost, depending on how you use the system and what type of climate you live in. The same subsidies that apply to solar photovoltaics apply to solar water heaters, so the cost can be reduced significantly. At this writing, a federal tax credit of 30 percent applies to both solar hot water and solar PV. That's a big incentive.

And if you install a solar water heater, you can install solar PV later on; they work very well together. You can make a smaller investment now, and leave the big bucks for later (when you get that raise you've been working so hard for). Just make sure you plan ahead and have enough roof space because solar PV is fussier about how it's placed on a roof. If you want to install solar hot water, followed by solar PV a few years later, get quotes and information for both types of systems now and you'll understand the limitations and advantages so that you can get the most benefit.

Consult my book *Solar Power Your Home For Dummies* (Wiley) for a lot more details on how to use solar power.

Using Biofuels to Power Your Car

You may not even know it, but most new cars can burn biofuels, like ethanol, without your making any changes to the engine. You can find local gas stations that provide biofuels by asking around or consulting the Internet, and you can make a habit of filling up at these stations.

You can also make some minor modifications to your auto's engine and burn higher concentrations of biofuels, or different types of biofuels. See Chapter 17 for more details.

Biofuels burn cleaner and the engine works better. Plus the carbon cycle is closed loop so you're not contributing to global warming as much as if you use pure gasoline. You won't be able to tell the difference in your car's performance either.

Driving a Hybrid

The cost of a hybrid auto or truck is higher than the cost of an equivalent gasoline-powered vehicle, but you'll get much better gas mileage, and you'll be emitting a lot less pollution. So while the upfront cost may be more, you'll see savings all along the way in terms of paying for gas. See Chapter 19 for more details on driving hybrids.

Hybrids are here now, and there are more and more of them on the road every year. Most major auto manufacturers offer various versions of hybrids, so your choices are broad, and getting more so every year. Many auto manufacturers offer their conventionally powered autos with hybrid options (for instance, Camry, by Toyota). You won't be able to tell the difference in performance, but your gasoline bills will be much smaller.

There's a less direct effect as well. When you buy a hybrid, you help reduce the cost of future hybrids so that even more people will get into the game. The cost of any manufactured product is a function of how many of them are built, so the more hybrids on the road, the less they will cost. When you buy a hybrid, you're investing in the future in both a direct and indirect way.

Installing a Geothermal Heating and Cooling System in Your Home

I explain the pros and cons of geothermal energy in Chapter 12. The technology is not cheap, but it's extremely clean and reliable. New homes, in particular, can benefit from the installation of geothermal because it's much less expensive to start a geothermal system from scratch than to retrofit an existing home.

You will need to have access to the right soil conditions, but if you do, this option is an excellent investment, particularly where heating and cooling bills are high year-round. The upfront cost of geothermal equipment is high, but the power bills will be forever low. You'll not only get lower power bills, but you'll be immune from energy price spikes in the future. The value of a home equipped with geothermal equipment is much higher than a conventional home, so you'll recoup your investment if you decide to sell your home. The higher energy prices go, the more your geothermal system will be worth.

Driving with Diesel (Without the Stink)

In Europe, diesel autos and trucks are numerous and common. Diesel-powered propulsion is more efficient than gasoline, and with the right equipment, the pollution levels can be much lower as well.

The diesel vehicles that were sold in the United States back in the early '80s gave diesel an unwarranted bad name, and most Americans equate diesel with dirty, sooty exhaust and strange sounds. New diesel vehicles, with European-developed technology, are clean, powerful, and efficient.

Diesel fuel is available almost everywhere, although sometimes it costs a little more. But because you'll be getting better gas mileage, your fuel bills will be lower, on net. Diesel engines last longer than gasoline-powered engines and require a lot less maintenance.

Best of all, use biodiesel (See Chapter 17 for more details). It's hard to find right now, but that'll change as more people get into the game. Diesel engines can burn a wide range of different fuel sources without need for modification.

Installing a Windmill in Your Own Backyard

Wind power is going to grow exponentially in the next few decades simply because it makes a tremendous amount of sense. If you have the right conditions, a small-scale windmill can give you a return on investment much better than most other alternative-energy options. In particular, the payback period for a good windmill is around two thirds that of a solar PV system. Wind is generally available more often than solar power, which only works when the sun is shining.

You'll need a windy location, and you'll need to consider such things as sound and how the windmill affects the scenery. You may have to get special permits, so consult your county building department. You may also want to check with your neighbors or your neighborhood association, if you have one, as they will be sharing the sound and visual effects with you, whether they want to or not.

You can install a small-scale windmill yourself, but there are some complications that require you to be very good with tools and hardware systems. If you don't install a windmill properly, it can be very dangerous. Check out Chapter 11 for more details.

Moving into a Smaller, Energy-Efficient Home

In general, the smaller the home, the less energy the home consumes. Many large homes have long hallways and awkward layouts that are not conducive to maximizing space. A good small home can often feel every bit as spacious as a good-sized mansion, if the design is done well. My personal impression of large homes is that they are more for showing off wealth than creature comfort. That sort of attitude is becoming obsolete in this day and age of energy concern.

If you're buying a new home, or an existing home, consider the layout, and the way the existing space is partitioned. Also consider how energy efficient the home is. Look for energy-efficient appliances, and if possible, find a home with solar power, or geothermal. A good stove is a plus. You can ask for a history of power bills, but you'll have to determine how your own power consumption relates to that of the seller.

Living in an Urban Environment

Living in the suburbs entails long drives and big, autonomous homes, along with spacious yards and landscaping. All of this consumes inordinate amounts of energy. If you live in an urban environment, however, you can take advantage of mass transit and the amount of energy required for modular and high-density living is much less than for a home sitting all by itself.

Government studies have indicated that urban living consumes around 60 percent of the energy, per capita, that suburban or rural living does. That also equates to around 60 percent of the carbon footprint. Urban buildings are generally fitted with natural gas pipelines, which are difficult to find in most suburban and rural environments. In addition, many urbanites don't even own cars, which equates to big energy savings. And smaller cars are more practical in cities as well. The average auto mileage for vehicles owned by people who live in cities is around 25 percent higher than for vehicles owned by suburbanites.

Investing in Energy Stocks

There are thousands of energy companies, some better than others. Many new startups are dedicated to developing and selling green technologies, for instance solar and wind power companies.

Instead of insisting on a maximum return from your investment dollars as your only criteria for investment, you can also look for companies that promote environmentalism. You can even find companies that are green, and also make a lot of green. Energy is a hot topic these days, and the buzz itself often creates profitable ventures. However, be sure to do your research and make sure your "green" company is truly green because some unscrupulous companies claim undue environmental concern.

Wind power companies, in my view, are going to grow quite a bit in the next few decades. If you're interested in a long term investment with a lot of green impact, find a good wind power company and buy its stock.

Driving Less

Perhaps this is too obvious to merit inclusion, but one of the biggest things you can do to help out with our energy problem is simply to drive less. Plan ahead; when you need to run errands, make a route that takes you to all the places you need to go in the most time-saving way. Take your vacations closer to home. Walk to those places you can reach on foot rather than drive. Carpool or use mass transit whenever possible. The list of ways to reduce your driving is endless.

Index

• D •

• F •

BUSINESS, CAREERS & PERSONAL FINANCE

Accounting For Dummies, 4th Edition*
978-0-470-24600-9

Bookkeeping Workbook For Dummies†
978-0-470-16983-4

Commodities For Dummies
978-0-470-04928-0

Doing Business in China For Dummies
978-0-470-04929-7

E-Mail Marketing For Dummies
978-0-470-19087-6

Job Interviews For Dummies, 3rd Edition*†
978-0-470-17748-8

Personal Finance Workbook For Dummies*†
978-0-470-09933-9

Real Estate License Exams For Dummies
978-0-7645-7623-2

Six Sigma For Dummies
978-0-7645-6798-8

Small Business Kit For Dummies, 2nd Edition*†
978-0-7645-5984-6

Telephone Sales For Dummies
978-0-470-16836-3

BUSINESS PRODUCTIVITY & MICROSOFT OFFICE

Access 2007 For Dummies
978-0-470-03649-5

Excel 2007 For Dummies
978-0-470-03737-9

Office 2007 For Dummies
978-0-470-00923-9

Outlook 2007 For Dummies
978-0-470-03830-7

PowerPoint 2007 For Dummies
978-0-470-04059-1

Project 2007 For Dummies
978-0-470-03651-8

QuickBooks 2008 For Dummies
978-0-470-18470-7

Quicken 2008 For Dummies
978-0-470-17473-9

Salesforce.com For Dummies, 2nd Edition
978-0-470-04893-1

Word 2007 For Dummies
978-0-470-03658-7

EDUCATION, HISTORY, REFERENCE & TEST PREPARATION

African American History For Dummies
978-0-7645-5469-8

Algebra For Dummies
978-0-7645-5325-7

Algebra Workbook For Dummies
978-0-7645-8467-1

Art History For Dummies
978-0-470-09910-0

ASVAB For Dummies, 2nd Edition
978-0-470-10671-6

British Military History For Dummies
978-0-470-03213-8

Calculus For Dummies
978-0-7645-2498-1

Canadian History For Dummies, 2nd Edition
978-0-470-83656-9

Geometry Workbook For Dummies
978-0-471-79940-5

The SAT I For Dummies, 6th Edition
978-0-7645-7193-0

Series 7 Exam For Dummies
978-0-470-09932-2

World History For Dummies
978-0-7645-5242-7

FOOD, GARDEN, HOBBIES & HOME

Bridge For Dummies, 2nd Edition
978-0-471-92426-5

Coin Collecting For Dummies, 2nd Edition
978-0-470-22275-1

Cooking Basics For Dummies, 3rd Edition
978-0-7645-7206-7

Drawing For Dummies
978-0-7645-5476-6

Etiquette For Dummies, 2nd Edition
978-0-470-10672-3

Gardening Basics For Dummies*†
978-0-470-03749-2

Knitting Patterns For Dummies
978-0-470-04556-5

Living Gluten-Free For Dummies†
978-0-471-77383-2

Painting Do-It-Yourself For Dummies
978-0-470-17533-0

HEALTH, SELF HELP, PARENTING & PETS

Anger Management For Dummies
978-0-470-03715-7

Anxiety & Depression Workbook For Dummies
978-0-7645-9793-0

Dieting For Dummies, 2nd Edition
978-0-7645-4149-0

Dog Training For Dummies, 2nd Edition
978-0-7645-8418-3

Horseback Riding For Dummies
978-0-470-09719-9

Infertility For Dummies†
978-0-470-11518-3

Meditation For Dummies with CD-ROM, 2nd Edition
978-0-471-77774-8

Post-Traumatic Stress Disorder For Dummies
978-0-470-04922-8

Puppies For Dummies, 2nd Edition
978-0-470-03717-1

Thyroid For Dummies, 2nd Edition†
978-0-471-78755-6

Type 1 Diabetes For Dummies*†
978-0-470-17811-9

* Separate Canadian edition also available
† Separate U.K. edition also available

Available wherever books are sold. For more information or to order direct: U.S. customers visit www.dummies.com or call 1-877-762-2974.
U.K. customers visit www.wileyeurope.com or call (0)1243 843291. Canadian customers visit www.wiley.ca or call 1-800-567-4797.

 **WILEY**

INTERNET & DIGITAL MEDIA

AdWords For Dummies
978-0-470-15252-2

Blogging For Dummies, 2nd Edition
978-0-470-23017-6

Digital Photography All-in-One Desk Reference For Dummies, 3rd Edition
978-0-470-03743-0

Digital Photography For Dummies, 5th Edition
978-0-7645-9802-9

Digital SLR Cameras & Photography For Dummies, 2nd Edition
978-0-470-14927-0

eBay Business All-in-One Desk Reference For Dummies
978-0-7645-8438-1

eBay For Dummies, 5th Edition*
978-0-470-04529-9

eBay Listings That Sell For Dummies
978-0-471-78912-3

Facebook For Dummies
978-0-470-26273-3

The Internet For Dummies, 11th Edition
978-0-470-12174-0

Investing Online For Dummies, 5th Edition
978-0-7645-8456-5

iPod & iTunes For Dummies, 5th Edition
978-0-470-17474-6

MySpace For Dummies
978-0-470-09529-4

Podcasting For Dummies
978-0-471-74898-4

Search Engine Optimization For Dummies, 2nd Edition
978-0-471-97998-2

Second Life For Dummies
978-0-470-18025-9

Starting an eBay Business For Dummies, 3rd Edition†
978-0-470-14924-9

GRAPHICS, DESIGN & WEB DEVELOPMENT

Adobe Creative Suite 3 Design Premium All-in-One Desk Reference For Dummies
978-0-470-11724-8

Adobe Web Suite CS3 All-in-One Desk Reference For Dummies
978-0-470-12099-6

AutoCAD 2008 For Dummies
978-0-470-11650-0

Building a Web Site For Dummies, 3rd Edition
978-0-470-14928-7

Creating Web Pages All-in-One Desk Reference For Dummies, 3rd Edition
978-0-470-09629-1

Creating Web Pages For Dummies, 8th Edition
978-0-470-08030-6

Dreamweaver CS3 For Dummies
978-0-470-11490-2

Flash CS3 For Dummies
978-0-470-12100-9

Google SketchUp For Dummies
978-0-470-13744-4

InDesign CS3 For Dummies
978-0-470-11865-8

Photoshop CS3 All-in-One Desk Reference For Dummies
978-0-470-11195-6

Photoshop CS3 For Dummies
978-0-470-11193-2

Photoshop Elements 5 For Dummies
978-0-470-09810-3

SolidWorks For Dummies
978-0-7645-9555-4

Visio 2007 For Dummies
978-0-470-08983-5

Web Design For Dummies, 2nd Edition
978-0-471-78117-2

Web Sites Do-It-Yourself For Dummies
978-0-470-16903-2

Web Stores Do-It-Yourself For Dummies
978-0-470-17443-2

LANGUAGES, RELIGION & SPIRITUALITY

Arabic For Dummies
978-0-471-77270-5

Chinese For Dummies, Audio Set
978-0-470-12766-7

French For Dummies
978-0-7645-5193-2

German For Dummies
978-0-7645-5195-6

Hebrew For Dummies
978-0-7645-5489-6

Ingles Para Dummies
978-0-7645-5427-8

Italian For Dummies, Audio Set
978-0-470-09586-7

Italian Verbs For Dummies
978-0-471-77389-4

Japanese For Dummies
978-0-7645-5429-2

Latin For Dummies
978-0-7645-5431-5

Portuguese For Dummies
978-0-471-78738-9

Russian For Dummies
978-0-471-78001-4

Spanish Phrases For Dummies
978-0-7645-7204-3

Spanish For Dummies
978-0-7645-5194-9

Spanish For Dummies, Audio Set
978-0-470-09585-0

The Bible For Dummies
978-0-7645-5296-0

Catholicism For Dummies
978-0-7645-5391-2

The Historical Jesus For Dummies
978-0-470-16785-4

Islam For Dummies
978-0-7645-5503-9

Spirituality For Dummies, 2nd Edition
978-0-470-19142-2

NETWORKING AND PROGRAMMING

ASP.NET 3.5 For Dummies
978-0-470-19592-5

C# 2008 For Dummies
978-0-470-19109-5

Hacking For Dummies, 2nd Edition
978-0-470-05235-8

Home Networking For Dummies, 4th Edition
978-0-470-11806-1

Java For Dummies, 4th Edition
978-0-470-08716-9

Microsoft® SQL Server™ 2008 All-in-One Desk Reference For Dummies
978-0-470-17954-3

Networking All-in-One Desk Reference For Dummies, 2nd Edition
978-0-7645-9939-2

Networking For Dummies, 8th Edition
978-0-470-05620-2

SharePoint 2007 For Dummies
978-0-470-09941-4

Wireless Home Networking For Dummies, 2nd Edition
978-0-471-74940-0

OPERATING SYSTEMS & COMPUTER BASICS

iMac For Dummies, 5th Edition
978-0-7645-8458-9

Laptops For Dummies, 2nd Edition
978-0-470-05432-1

Linux For Dummies, 8th Edition
978-0-470-11649-4

MacBook For Dummies
978-0-470-04859-7

Mac OS X Leopard All-in-One Desk Reference For Dummies
978-0-470-05434-5

Mac OS X Leopard For Dummies
978-0-470-05433-8

Macs For Dummies, 9th Edition
978-0-470-04849-8

PCs For Dummies, 11th Edition
978-0-470-13728-4

Windows® Home Server For Dummies
978-0-470-18592-6

Windows Server 2008 For Dummies
978-0-470-18043-3

Windows Vista All-in-One Desk Reference For Dummies
978-0-471-74941-7

Windows Vista For Dummies
978-0-471-75421-3

Windows Vista Security For Dummies
978-0-470-11805-4

SPORTS, FITNESS & MUSIC

Coaching Hockey For Dummies
978-0-470-83685-9

Coaching Soccer For Dummies
978-0-471-77381-8

Fitness For Dummies, 3rd Edition
978-0-7645-7851-9

Football For Dummies, 3rd Edition
978-0-470-12536-6

GarageBand For Dummies
978-0-7645-7323-1

Golf For Dummies, 3rd Edition
978-0-471-76871-5

Guitar For Dummies, 2nd Edition
978-0-7645-9904-0

Home Recording For Musicians For Dummies, 2nd Edition
978-0-7645-8884-6

iPod & iTunes For Dummies, 5th Edition
978-0-470-17474-6

Music Theory For Dummies
978-0-7645-7838-0

Stretching For Dummies
978-0-470-06741-3

Get smart @ dummies.com®

- **Find a full list of Dummies titles**
- **Look into loads of FREE on-site articles**
- **Sign up for FREE eTips e-mailed to you weekly**
- **See what other products carry the Dummies name**
- **Shop directly from the Dummies bookstore**
- **Enter to win new prizes every month!**

Do More with Dummies

Products for the Rest of Us!

Acrylic Painting Set FOR DUMMIES

Embroidery Broderie DUMMIES

Sewing Patterns DUMMIES

Wall & Ceiling Repair Kit

No More Slice Kit DUMMIES

OrganizeMY Electronic Filing DUMMIES

Smart Booster Cables

GPS Navigation DUMMIES

Jigsaw Puzzle DUMMIES

Cables DUMMIES HDTV Cable Kit

...ure • Crafts

...nore!

Check out the Dummies Product Shop at www.dummies.com for more information!

WILEY